THOMAS BERRY

THE SACRED UNIVERSE

EARTH, SPIRITUALITY, AND RELIGION IN THE TWENTY-FIRST CENTURY

Edited and with a Foreword by Mary Evelyn Tucker

Columbia University Press New York

Columbia University Press
Publishers Since 1893
New York Chichester, West Sussex

Library of Congress Cataloging-in-Publication Data
Berry, Thomas Mary, 1914–
The sacred universe : earth, spirituality, and religion
in the twenty-first century / Thomas Berry ;
edited and with a foreword by Mary Evelyn Tucker.
p. cm.
Includes bibliographical references (p.) and index.
ISBN 978-0-231-14952-5 (cloth) —
ISBN 978-0-231-52064-5 (e-book)
1. Spirituality. 2. Religion. 3. Religions. 4. Cosmology.
5. Evolution. 6. Philosophical anthropology. I. Title

BL624.B4638 2009
200.9´05—dc22 2009005329

Columbia University Press books are printed on permanent
and durable acid-free paper.
This book is printed on paper with recycled content.
Printed in the United States of America
c 10 9 8 7 6 5 4 3 2 1

To Fanny and Ted deBary
in celebration of sixty years of friendship

Contents

PART IV

Foreword

Mary Evelyn Tucker

I REMEMBER the first time I met Thomas Berry, on a cold winter day in early February 1975. He was living along the Hudson River just north of New York City at his Riverdale Center for Religious Research. We sat in his sun porch overlooking the Palisades and under the spreading branches of the great red oak.

I had just returned from two years in Japan, where I was teaching at a university in a provincial capital five hundred miles south of Tokyo. It was, indeed, another world. I had traveled through Asia on my way back home, encountering Taiwan, Hong Kong, and Singapore in their early phase of economic development. I also traveled to Vietnam, close to the end of the war, and to India, which was still struggling to feed its burgeoning population.

The immense inequities, the complex histories, and the vastly different religions were overwhelming to me—so recently out of college and so woefully ignorant of Asia. I needed to sort out this disorienting experience. I came to Thomas Berry because he was immersed in studying the religious traditions of Asia. He had created a successful graduate program in the history of religions at Fordham University along with his research center in Riverdale. He had published a book on Buddhism and one on the religions of India. I knew I could learn something here.

I had written to Thomas while I was teaching in Japan, because I had been given some of his unpublished essays by Fanny de Bary before I left. She and her husband, Ted de Bary, a scholar of Confucianism at Columbia University, were two of Thomas's closest friends. I was amazed to discover this remarkably balanced perspective, which neither idealized "Eastern religions" nor moved away from "Western religions," as was fashionable then. Berry avoided such simple interpretations, seeking rather to discover the wellspring of wisdom in each of these traditions. After studying their texts and examining their practices, he wrote insightful essays comparing and contrasting these religions with a judicious eye and a rich sense of historical complexity.

In those days, Berry mimeographed his essays, then collected and bound them together in what he called the "Riverdale Papers." In a period before computers or e-mail, they circulated among interested readers in the New York area and beyond. Some of those early essays are published at the opening of this book and reflect the sweeping intellectual vision of Berry in his sixties. The middle essays were written on his first computer, which he purchased and learned to use in his seventies. The concluding essays were written in his early eighties, when he was moving away from Riverdale and back to North Carolina.

These essays, written over three decades, reflect the remarkable journey of Thomas Berry to encompass ever-widening circles of history. Beginning as a historian of Europe, immersed in its religious, cultural, and intellectual currents, Berry expanded this approach to embrace Asia as well. He read widely in world history and religion and built a library of some ten thousand books at Riverdale. This reading culminated in his move beyond human history to Earth history, as he traced his way back into deep time. He was especially influenced by the integrating evolutionary vision of Pierre Teilhard de Chardin.

Knowing on some intuitive level that we humans are guided by story, he ultimately called for the telling of the universe story. He felt that it was only in such a comprehensive scale that could we situate ourselves fully. His great desire was to see where we have come from and where we are going amid ecological destruction and social ferment. It was certainly an innovative idea, to announce the need for a new story that integrated the scientific understanding of evolution with its significance for humans. This is what he found so appealing in Teilhard's seminal work *The Human Phenomenon.*

But how did this all begin? What was Berry's own evolution of ideas and influences? As a young graduate student, he was already striving to see the whole. He wrote his Ph.D. thesis on Giambattista Vico, exploring his philosophy of history, which posited three stages in world history, namely, the age of the gods, the age of the heroes, and the age of humans, as well as the significant "barbarism of reflection" that ended each age and gave rise to the next. In later years, he explored other historical perspectives, such as those of Herbert Spencer, Oswald Spengler, Arnold Toynbee, Christopher Dawson, Eric Vogelin, and William McNeill.

But first he studied world religious traditions, beginning with his own Christianity. He read the Church Fathers and Augustine and the medieval theologians, including Bonaventure and especially Thomas Aquinas. He explored such mystics as John of the Cross, Teresa of Avila, and Meister Eckhart, and literature associated with the rise of the Beghards and the Beguines in the Low Countries. He pored over the twentieth-century century debates regarding modernism and the Second Vatican Council. He was seeking a way to understand the power of tradition and the challenge of modernity. How Christianity and other religions changed and developed over time became a major focus of his thinking. As a historian, he was fascinated by the rhythms of continuity and change displayed in religions.

As his studies spread to other religions, he began to identify some of the comparative patterns of religious development from the initial classical period and the appearance of great religious leaders such as the prophets of Israel, Buddha in India, or Confucius in China. With the early growth and spread of traditions came various theologies or schools of thought, which began to differentiate themselves, such as Theravada and Mahayana Buddhism. In the Middle Ages, new syntheses arose with such comprehensive thinkers as Thomas Aquinas in Christianity, Moses Maimonides in Judaism, Al Ghazali in Islam, Chu Hsi in Confucianism, and Shankara in Hinduism.

In many of these cases, the new synthesis was created in response to the challenge of other religions, for example, Buddhism for Confucianism and Islam for Christianity. These new formulations became effective containers of the rich symbolic and ritual life of cultures around the world and inspiring visions for the spiritual yearnings of modern humans. Berry studied these historical patterns of religious development and began to focus on the challenge that the modern period presented to religious traditions. In addition, he became increasingly concerned with the effects of rapid industrialization on the ecosystems of the planet and the lack of response of the religions to this growing crisis.

Berry was profoundly aware of the deep alienation that had beset twentieth-century civilization, torn as it was by two world wars and confronted with an existential crisis of meaning. He recognized that religions and their larger spiritual visions have something of immense significance to offer humans in their struggle to manage the demands and tragedies life presents. While recognizing the shortcomings of the various religions, he nonetheless creates space in these essays for religions to open up to modernity—to reinhabit their symbol systems and enlarge their embrace so as to grapple with modern culture and thought.

In addition, Berry urges the religious traditions into a deepening reflection on the encounters with one another that an age of global communications and travel has made possible. He calls for a dialogue of civilizations and religions, not a clash—and this more than thirty years ago. His own close encounter and study of other religions began with his time in China in 1948 and 1949, where he became fascinated by Confucianism as well as by Daoism and Buddhism. With Ted de Bary he chaired the Oriental Thought and Religion seminar at Columbia in the years after they both returned from China. Here the study of Asian religions and their encounter with the West was fostered.

At one of these seminars, Berry presented a paper on Buddhism, and his penetrating insights into other traditions became clearly visible. He spoke on the Shingon sect, the esoteric or Tantric form of Japanese Buddhism that relies heavily on the art of the mandala. Professor de Bary noticed that one of the attending professors, Yoshito

Hakeda, who had gone through the training to become a monk in the tradition, was silently weeping. When later de Bary asked him why he wept, Hakeda said that in hearing Berry he felt as though he was hearing his Buddhist teacher on Mt. Koya in Japan.

Berry often moved his students with his unusual ability to understand the spiritual depths of other religions, even though the concepts and terminology were so very different. He could explain these spiritual ideas and practices with clarity for the uninitiated and with freshness even for those familiar with the tradition. His study included not only the religious traditions but also the culture and arts of Asia. He had a large collection of books on Asian poetry, literature, painting, and sculpture. All of this was considered part of understanding a religious tradition from his perspective. Berry's invitation for the religions to review and expand their symbol systems comes out of a deep appreciation of religious art and culture. In addition, his call to be in dialogue with other religions arose from his understanding that religions have been continually changing and adapting in their contacts with one another. This is becoming even more pronounced in the modern period.

These essays represent an evolution in Berry's thinking over some three decades of teaching and lecturing. They reflect the broad research and reading that Berry was engaged in for many years—during his graduate studies, his year in China, his time in Europe, his teaching at St. John's and Fordham Universities, and his twenty years at the Riverdale Center. Most of these essays were written while he directed the Riverdale Center from 1973 to 1993 and while he served as president of the American Teilhard Association from 1975 through 1987. These were years of enormous creativity, with lecture series in the fall and spring and a workshop each summer. Berry was surrounded by his graduate students from Fordham and Columbia and by many intellectuals and seekers from the New York area and beyond.

They came to hear him speak because he was a wellspring of probing questions, fresh angles of vision, and provocative ideas. Most of all, one could sense he was finding a way forward out of the alienation of modern industrialized life and beyond the stale forms of religious life, out of the fragmentation of social communities and into a more integrating and comprehensive framework.

This framework included the opening up of the religions to restore and articulate anew their vital, nourishing messages. He wished to bring the continuity of their historical tradition into conversation with the changing nature of the modern world. This involved both renewal within and dialogue without. It required dialogue with other religions, dialogue with science, and dialogue with the world at large. Most especially, the looming ecological and social crisis was becoming more visible during those years, and Berry's brooding mind and empathetic sensibility sought to understand how to respond.

And so we can trace in the first four essays a call for the religions to open up, to become healers of alienation, to become beacons of inspiration for the religious imagination. He recognized that a comprehensive spirituality is being established, for we are now inheriting the global human heritage of the world's religions.

For Berry, this requires an expansion of space to inhabit both our bioregion and the planet as a whole. We are part of a multicultural, planetary civilization. It also demands an extension in time, so that we deepen our temporal historical awareness. We are biohistorical beings living in the context of universe and Earth energies.

In part 2, we see Berry's perceptive call several decades ago to recognize that human spirituality is primarily related to the spirituality of the Earth. He observes that the overemphasis on redemption in contrast to creation has left many people alienated from the wonder and beauty of the Earth and universe. This inability to interact creatively with the larger Earth community—its air, water, soil, its species and ecosystems—has resulted in immense loss and destruction. For Berry, religion in the twenty-first century needs to recover this sense of the manifestation of the divine in the natural world. In recognizing this, we can create religious symbols and rituals that will truly orient us to our cosmological being.

In part 3, Berry calls for religions to recover their cosmological sensibilities, to see the human as microcosm as profoundly related to the universe as macrocosm. He explores the Gaia hypothesis, which illustrates the homeostatic self-regulatory nature of Earth. He observes that we are growing in our understanding that the study of life is integral to the story of the Earth itself. For Berry, this is a sacred story that has emerged with great creativity. We awaken to this creativity with wonder and awe, which Berry sees as profoundly religious sensibilities.

In the final part, Berry has arrived more fully at this central message. We dwell in a sacred universe, we are part of a vast evolving process, we are returning to a sense of kinship with all beings. Most especially, Berry observes that a primary locus of meeting the divine is in the natural world. To extinguish a species is to extinguish a voice of the divine.

Berry brings us back to the North American continent in a powerful evocation of the beauty and grandeur of life that has inhabited this land. He ponders the causes of the loss of so much of this biodiversity and seeks new ways for a viable future. In particular, he calls for the awakening of wonder, so that along with such new strategies as sustainable agriculture, ecological economics, green politics, and ecodesign there will also be an emerging sensibility in human consciousness that will have the enduring energy for the great transition ahead.

This is Berry's fondest hope—that the dynamizing sources of human energy will be found in a broadened religious and spiritual sensibility. This comprehensive sensibility includes a revitalization of the world's religions and a robust dialogue among and between civilizations. It also invites an opening to the universe story as a context for situating our biohistorical nature in a multireligious planetary culture of peace.

There are few individuals who have arrived at such a comprehensive and inclusive vision with prolonged and careful study—one affirming both differentiation between and communion among peoples and religions. Berry's distinctive contribution is that he did not stop there; he also gave to this moment of global encounter among religions and civilizations a context in which to understand our common origins in evolutionary processes and thus to recognize the need to create a shared future with all beings in the Earth community.

THE SACRED UNIVERSE

PART I

CHAPTER 1

Traditional Religion in the Modern World

THE MODERN scientific world arose out of a past powerfully influenced by religion. Yet from the sixteenth century there has been tension between our religious traditions and our modern scientific modes of understanding. A basic antagonism has existed on both sides, along with a limited amount of appreciation, approval, and mutual support. In general, however, it can be said that relations between the two in modern times have never been adequately managed or appreciated. This seems to be one of the special tasks to which our present generation is called.

What we look for is not a total understanding or agreement between religious traditions and scientific developments but a mutually supportive relationship and appreciation. Each must remain in some manner a mystery to the other, as love and reason are mysteries to each other, although each functions most effectively when it modifies and supports the other. The proper integration of the diverse dimensions of human life requires neither total integration nor mutual exclusion; rather, it consists of an interplay of tensions that are both functional and creative.

This conflict of modern and traditional values was originally a problem for the Western world but now has become a worldwide problem.

Traditional civilizations are experiencing a certain threat in the face of the modern, aggressive, demanding, secular world. While there are many aspects of these problems that deserve attention, I present three for consideration.

First, the modern, scientific, technological world is not primarily responsible for the contemporary decline in religious life. Second, science is indebted to the Western religious traditions. Third, modern thinking and Western religious thinking are committed to history as a development process. Religious life in its traditional forms, including its power as a creative force in the cultural life of traditional civilizations, reached a certain state of stagnation prior to the rise of our scientific, secularist world. In the Western Christian world, if we examine the human situation as far back as the fifteenth century, we find that the shock of the plague, the Black Death, produced a negative attitude in the Christian orientation toward the universe and the natural life systems, especially the body.

Even earlier, in the 1350s, the Dominican preacher Jacopo Passavanti (1302–1357) spoke vividly in the Santa Cruz church of Florence of the pains of hell and the need for a severe spiritual regime to negate the body and the world so as to turn toward the heavenly realm. In the following century, the Augustinian friar Thomas à Kempis (1380–1471) wrote the influential treatise *The Imitation of Christ*, in which he proposed a spiritual orientation based on detachment from the concerns of everyday life. Throughout the fifteenth century, an intense devotionalism detached from worldly concerns was taught as the basic mode of Christian spirituality. This devotionalism found expression in German theology, which led to the gospel purism of the devotional movements.

This contributed to the religious difficulties of sixteenth-century Protestant Puritanism and later Catholic Jansenism of the seventeenth century. Ultimately, the antipathy between Catholics and Protestants in Europe resulted in the devastating Thirty Years War (1618–1648). Published in 1772, *La Grande Encyclopedie* defined the secular, scientific, Enlightenment period that has continued substantially into the present. Christianity had entered one of its least satisfactory states at the end of the eighteenth century, a situation that continued throughout the nineteenth century in both Catholic and Protestant traditions. Judaism also struggled to keep its authentic expression amid the powerful influences of modern movements in central Europe. The Hasidic movement in Poland, which originated with the teachings of Baal-Shem-Tov (1698–1760), was among the most impressive efforts toward a new expression of Jewish authenticity at this time.

In North America, Christian groups were struggling to sustain traditional forms of Christian expression. They did maintain a certain institutional vigor. The difficulty was that the public presence of religion was progressively diminished in any effective form. In the United States, public institutions were more associated with Protestant Christianity.

When we turn from Europe and North America to Asia, especially to India and China, we find these vast civilizations in a period of self-reassessment. India at the end of the eighteenth century was in a difficult cultural situation. The ancient traditions of learning had declined, partly as a result of foreign political domination of large sections of the country and partly because of the inherent tensions within India's social structure. Other causes of decline were also at work—the most significant was a certain passivity in the face of life's difficulties, derived from the doctrine of karma.

Buddhism, founded by Gautama Buddha (560–480 bce), spread throughout India and moved to Sri Lanka in the third century bce, then to Southeast Asia. A later development of Mahayana Buddhism in the first century bce spread extensively throughout India, then to China in the first century ce and to Japan by the mid-sixth century ce. Tibet adopted Mahayana Buddhism in the sixth century ce. Although Buddhism prospered throughout most of Asia, it disappeared as a significant force within India, partly because of the Muslim invasions of the eleventh and twelfth centuries. Also, in the opening years of the second century ce, a new, more devotional Hinduism developed throughout India, evident in the early Puranas.

China was in a state of cultural equilibrium, and life continued in its fixed patterns. Since 1644, China had been ruled by a foreign dynasty, the Manchu. Much was achieved politically and in the realm of scholarship during the Ching period, yet the deeper creative forces of the spiritual, cultural, and intellectual orders were no longer expressing themselves with the vigor of earlier periods.

If we look to the world of Islam, we find a similar period of stagnation. There had been great political and cultural achievements in the sixteenth century, during the reigns of Suleiman the Magnificent (1520–1566) and Akbar in India (1556–1605). But the spiritual order of the Islamic world of the late eighteenth century was clearly less creative than in earlier periods.

In the mid-sixteenth century in the West, there was an amazing development, something on the cultural scale comparable to a violent eruption on the geological scale: the birth of modern science, fueled by a new conception of the universe initially set forth in a publication of Nicolaus Copernicus (1473–1543) in 1543. He was followed by Johannes Kepler (1571–1630), who discovered the elliptical movement of the planets. Then in the seventeenth century came René Descartes (1596–1650), Francis Bacon (1561–1626), Galileo Galilei (1564–1642), and Isaac Newton (1642–1727). A new experience of the universe had a shattering effect on every aspect of the existing Western culture.

The creativity of the West was now situated primarily in the scientific inquiry into the physical structure and biological functioning of the universe, the shaping of modern political nationalism, and the vast surge in the commercial-industrial world. If we survey the spiritual situation or state of religion in traditional civilizations at the beginning of the nineteenth century, we find little capacity on their part to provide the new spiritual interpretation needed to address the scientific-secular developments. Instead, spiritual traditions began to blame their own difficulties on the rising forces that directed these societies and provided a new set of secular values. This tension between the spiritual and the scientific continues into the present. The secular world blamed the surviving religious elements of Western life for the difficulties it was experiencing. Religions blamed secularism for the weakening of dedication to religious values. In both instances, the antagonism was founded on an effort to make the other a scapegoat for the deficiencies found in each.

My second proposition is, the tension just described notwithstanding, that scientific endeavor is profoundly indebted to the religious traditions of the West. This is because the religious traditions established the necessary conditions in which science could develop to its present state. The history of the scientific tradition, beginning in classical times, indicates that science depended greatly on the spiritual traditions of the West, in addition to Greek philosophy. In the ninth through twelfth centuries, Islam was more advanced in its intellectual development than the Christian West. Islam was even the teacher of Europe. Yet despite this high achievement, the Islamic spiritual traditions did not permit the development of a modern order of intellectual life. The intellectual life of Islam lost its creative power under a religious critique that could not reconcile reason and faith, the philosopher and the believer, and the inherent functioning of the natural world with divine direction of the universe. Commitment to rational thinking was regarded as the cause of the weakening of faith in the Qu'ran. The great age of Islamic thought in Spain, as exemplified by the Spanish Arab philosopher and physician Averrhoës (1126–1198), came to an end.

Meanwhile, the tension between faith and reason had been communicated to Christian Europe. Earlier, the thirteenth-century Christian thinkers had established a functioning relationship between reason and faith. They achieved this by recognizing the importance of both reason and faith and of the reality of the phenomenal world as well as the existence of a transcendent one. Thomas Aquinas (1224–1274) established a remarkable harmony between these poles. His emphasis on the reality of the phenomenal, rational, secular order; his intensive study of the logical processes of the human mind and the nature of the scientific endeavor; his insistence on the inherent efficacy of the secondary cause; and his statement that errors committed by human reason concerning the natural world led directly to errors concerning the divine world set up many of the conditions that permitted the birth of the modern world. Before Aquinas were Albert the Great (c. 1206–1280) and Roger Bacon (c. 1214–1294); after him came a great succession of individuals who carried the Western tradition into the brilliant scientific achievements that distinguish the modern human community. However, Thomas had so integrated traditional Christian thought with the deductive philosophical traditions of Aristotle that the synthesis he established made it difficult to understand or accept the new empirical inquiry into the functioning of the natural world.

Furthermore, Western Christianity, after establishing the context in which scientific development could take place, became alienated from this modern world by its failure to understand either itself or the world it had brought into existence. Anxiety over fidelity to its own heavenly origins weakened its ability to pursue a worldly course. In many ways, Christianity was experiencing difficulties born of its past success—and such difficulties are often less easily managed than outright failure.

Nonetheless, we should note certain fundamental religious orientations that conditioned Western society for the development of science. There is the commitment to the world as real and of supreme value. There is also Christianity's commitment to the powers of human reason and to the belief that reason and faith are mutually beneficial rather than mutually destructive. Christianity has paid such high tribute to human intelligence that it has often suffered from the accusation of confining its own higher spiritual vision within the narrow range of our rational powers of comprehension.

A third proposition concerning the relation of traditional Western religions to the modern world is that both are committed to history as a developmental process rather than simply as a cyclical renewal according to the norms of the cosmological time sequence of the solar calendar, especially the diurnal and seasonal sequences.

Here it is necessary to go beyond the question of the traditional and the modern and back to the most basic of all questions: the question of the human condition. The ultimate concerns of both science and religion are with the human condition in the full range of experience. Intellectually and spiritually, everything in human life depends on how we experience ourselves, how we respond to our life situation, and whether we manage the human condition in a creative or a destructive direction.

Whatever joy we may have in life, there is also a deep sense of tragedy built into our experience of ourselves and the world in which we exist. In its raw, uncultivated state, the human being is not satisfactory. The human condition is experienced as thoroughly and absolutely unsatisfactory. It must be altered to a degree so great that it is described as a new birth, a truly human and spiritual birth. Otherwise, the first birth never comes to term but is cut off in an undeveloped, savage condition. How to sustain the pain of existence meanwhile, how to give it meaning, then how to bring it under the influence of a transforming saving discipline: these are the basic challenges. Traditional religions consider that all the forces in heaven and Earth must contribute to this transforming process, to this new birth. This is the meaning of initiation rituals found among indigenous peoples, of the Hindu bestowal of the sacred cord, and of Christian baptism. This sense of giving a new birth to individuals and the community is the essential doctrine of Marxist socialism, as well.

In the traditional period, there was general agreement that this new birth brought us into a higher, sacred, or spiritual order that radiates over the whole of life and gives sublime meaning to every least detail of human existence. The larger purpose of life is to bring this spiritual birth to its full expression. It is not just salvation from the human condition—it is the transformation of the human condition itself.

Although this salvation doctrine of a higher spiritual birth is common to many traditions of Asia and the West, the West had from the beginning a unique awareness that salvation has a historical dimension. Humans attain this new birth only as members of the community in the course of its historical developmental context. Here we find the most profound agreement and also the greatest opposition between the traditional and the modern worlds. Increasingly, modern peoples are committed to historical communal salvation. There is, indeed, an extreme self-centeredness in modern humanity, a constant betrayal in favor of individual aggrandizement, but in so far as a person finds any commitment, it is to the community and to the more significant human process.

If we trace this sense of historical development back to its sources, we find that it was enunciated by the early prophets as the coming "Day of the Lord" toward which all temporal events were moving. This vision achieved clear expression in the teaching of the prophets Isaiah and Daniel. It later formed the conclusion of the Christian scriptures in the Apocalypse of John the Evangelist. This "Day of the Lord" is described as a period of peace, justice, and abundance. The very constitution of the world is to be altered. The lion and the lamb will lie down together. The nations will all come to the Mountain of the Lord. Swords will be beaten into plowshares. Heaven and Earth will be reconciled in a transformed world wherein the original human paradisal situation will be reestablished. All this is to be achieved in and through a transforming historical process.

I propose that nothing in Western life or in the life of the modern world can be understood in any depth apart from this historical vision, which originated in prophetic proclamation. It rang terribly clear in the Bolshevik revolutionary movement, which experienced itself as embodying the historical dynamic of the ages, as having as its mission the shattering of a past world to liberate us from the human condition and give a new birth in a higher order of a sacred classless community. This emphasis on a new birth process is clearly stated in the poets of Russia after the revolution of 1917. It stamps all forms of utopian revolutionary socialism of our times, whether legitimately or in distorted form, as expressing something deeply felt in our total religious life.

This same drive toward a new world is found in liberal democratic societies of the West. The sense of what this new birth is to be and how it is to be attained is vastly different, and that is what was so terrifying about the social, political, and military conflicts throughout the twentieth century. America was founded on the belief that this country was bringing about a new birth to humanity and that America, born out of a decadent European world, was the last great hope of the human community. A similar dedication to a transformation of the sociocultural development of the human community found expression in Marxist communism. Both democratic and Marxist societies agreed that the higher birth of humanity is an infrahistorical process—one within time and history—that has little to do with the spiritual rebirth into a higher transtemporal order of things as this was experienced by the earlier religious currents of the West.

This brings us to the main point of this discussion: the question of the earthly infrahistorical and the divine transhistorical orders. Are these to be considered separate, alien, and antagonistic to each other, or should they be considered as two phases of a single mode of being? It could be said that from the beginning there was a certain ambivalence in the scriptural pronouncements. It was not clear how much of the transformation was to be an individual, interior spiritual experience and how much was to be a transformation of the earthly conditions of human life. This has become the central ambivalence of our modern world. What is meant by salvation from the human condition? Is it merely an adjustment of human beings to the world of time and matter, or is it a higher spiritual process leading to a divine experience and liberating us from the confinements of temporal and spatial existence? Do we live simultaneously within a heavenly and an earthly kingdom?

The ideal of a transtemporal mystical experience of the divine remained the basic spiritual ideal of the Christian world until the late medieval period. Then the ideas of earthly progress began to take shape in the form of an infrahistorical, human, social, and scientific development over which we had a basic control. This sense of historical progress seized upon our world in a powerful way, especially through Hegel's sense of ontological development, Marx's sense of social development, Darwin's sense of physiological development, and Nietzsche's sense that the contemporary human is becoming a superior human type. With these ideas, the great religious task—the great religious experience—is no longer the ancient spiritual experience of divine presence, divine communion, or participation in divine life. Rather, it is the experience of an emerging humanity, of a new intellectual vision, of a new and more satisfying social order.

These are to be achieved through human social transformation and scientific, technological mastery of the surrounding world. Scientific and social ideals have both become substitute mysticisms. Technology is the sacrament of our new birth. With their inner mystical dimensions and outer efficacy, science and technology provide an analysis of the human condition and a transforming remedy. The infrahistorical drive of science must be thoroughly understood, for it is this new situation that constitutes the inner core of the question concerning the relationship between traditional spiritual values and the values of the modern world. Science is the main instrument for attaining the higher human development. Science, the world, human intelligence: these have become self-sufficient components of a new religious attitude.

Some maintain that this new secular city is the fulfillment of earlier traditions, which predicted the coming of a heavenly city. Is this the new, the true Christianity? To what extent is the eclipse of God and the neglect of ritual worship both a genuine humanism and a genuine form of religion? To what extent is this the authentic carrying out of the moral imperative of Genesis to transform the Earth, to develop and fructify the world?

Is this what it means to do the will of God on Earth as it is done in heaven? Is this to hallow the name of God? Is this to bring the heavenly kingdom to Earth? Is this the way to get our daily bread, to obtain forgiveness of our failures, to establish the human community, to be delivered from evil in its absolute form? If we answer affirmatively, then we can say that this age, in its structure and overall direction, is religious, even in a Christian sense.

Our problem, then, is to convert religion to the world rather than to convert the world to religion. Many are thinking in this direction, and with good reason. For whether we agree or disagree with the suggestion that religion should respond to the world, we must see the problem of the modern and traditional worlds in this context. Then we must ask: To what extent were traditional religious values bound up exclusively with a transtemporal, transearthly religious experience? Was not human religious rebirth always associated with a temporal-historical transformation process?

My answer would be that an integral understanding of Western religions, both Jewish and Christian, reveals that they have always had a commitment to the order of time, to our human, earthly destiny, and to social justice. The modern world, while extreme in its attitude, does not constitute any essential attack on the religious traditions of the West except where it misunderstands both itself and the religious tradition out of which it was born.

I am not saying that it is now possible to convince moderns of the religious aspects of their work or that it is possible to convince religious persons that this work in the order of time, society, and history is the fulfillment rather than a denial of traditional values. But whether either is convinced of the validity of the other position, the fact remains that this is our challenge. It demands our attention.

It should be observed that the world committed simply to the secular, scientific order does not find abiding satisfaction in what it is achieving. Ignorant of any spiritual significance in what we are doing, we remain profoundly dissatisfied, inwardly starved, spiritually and humanly debilitated, and unable to carry out successfully our finest endeavors. The world is not experiencing the higher rebirth that is needed. The changes effected by scientific and technological improvements of our earthly status are not sufficient. It may well be the work of the spiritual tradition of the West to reveal to the modern world its own nobility and to cure its self-distrust and reflexive cynicism at its failures. This is at least one essential aspect of the rebirth, this commitment to the higher human process. Christianity, however concerned with the experience of the divine, also considers this earthly task an integral part of the total creative-redemptive process.

But when we look closer, we see that this cynicism we moderns have for our own work is extremely difficult to deal with, because there is a basic cruelty in this higher human process. Its achievements are shared by an elite lifted higher and higher as time goes on, while the submerged masses of humanity enter into a new agony made the more bitter by our deep sensitivity to human suffering and by our awareness that new knowledge and powers are capable of relieving much of the agony to which humanity is presently condemned. Some of this agony, indeed, some of the deepest agonies in the slums of our cities around the world, is the direct result of "progress" itself. What is so frustrating is that efforts to produce remedies consistently fail in a manner that leads to a mood of desperation, unmitigated cynicism, violent outbursts, and destructive revolutionary movements. Some of these threaten the foundations of the existing social and political order. We are experiencing a rising anarchy in social relations that gives every thinking person forebodings about the future, even our immediate future.

Beyond this is the deeper tragedy of inner spiritual destitution that affects us even in the midst of our economic, intellectual, and social success. We find ourselves still deeply alienated, just when we have done all that we project for ourselves on the earthly and human plane. This is mirrored in the theater and literature of the absurd; here we find the revelation of humans as a despicable reality, ignoble even in our highest aspirations, disoriented, deteriorated beings. We are left with bleak visions such as those of Jean Genet (1910–1986), who openly revels in his world of perversion and depravity.

In analyzing this, we find that while humans have fulfilled a certain part of their lives, while they have dedicated themselves to high and noble causes, they still need a type of rebirth that can be provided only by some higher spiritual interpretation and inspiration of life. Jean Genet's and Samuel Beckett's characters cannot civilize, cannot constructively guide the destinies of humanity, and cannot inspire the loyalties needed to sustain a civilized, or even partially human, order of life.

Yet we can say of our modern world that in its success it is living off the capital of a former spiritual vision and is thus possibly squandering its inheritance. It might also be said that in its failures it is without inner spiritual significance. Nonetheless, it might be proposed that the modern experiences of social anarchy, deconstruction, and the radical absurdity of all existence are a preparatory phase, an effort at total honesty, a purifying of our illusions, and thus, at least, a beginning.

While all this is taking place, traditional religion, alienated from the modern world, has reached a spiritual impasse. It has lost much of its feeling for the genuine, for the authentic. It has shown neither the intelligence nor the willingness to walk with us through this modern period in our splendor and in our shame. Religion has not fully communicated the vital spiritual nourishment and illumination needed by a suffering world.

What, then, is needed? A modern world responsive to the spiritual—and spiritual traditions responsive to the modern world. These must be mutually infolded in the common task of bringing us to our full birth as human beings. I lift up for our consideration four modern spiritual personalities, from four distinct traditions, who have made significant contributions to this work.

The first is Aurobindo Ghose (1872–1950), intellectually one of the most creative of the spiritual personalities of modern India. In his work *The Life Divine*, he brings the historical aspect of Indian tradition into the modern world's sense of developmental time. He decisively answers those who say that Indian spirituality has nothing to contribute to modern humanity and insists that Indian spiritual traditions are fully capable of entering into the modern world of science and historical development. He shows the necessity of a spiritual interpretation of the higher human process if it is to have any final significance. He also indicates that the spiritual traditions of India can no longer exist in any vital way apart from these modern developments, which are themselves manifestations of higher human spiritual consciousness. What is important is the attainment of a conscious realization of the spiritual nature of human development. Only then can a truly integral human experience be achieved. The political and economic, the scientific and technological, the artistic and the literary, the spiritual and the religious are all valid forms of human existence that must be preserved, developed, and integrated with one another.

The second spiritual personality is Mohammed Iqbal (1873–1938), a poet-philosopher whose influence has spread throughout the Muslim world. Born in a Muslim region of India, he studied in Germany and at Cambridge, where he came under the influence of Henri Bergson (1859–1941) and Friedrich Nietzsche (1844–1900). Yet his deepest roots are in the Qu'ran and the Persian Islamic tradition, especially the high humanistic spirituality of Jalal-u'd-din Rumi (1207–1273). Iqbal awakened Islam from its mystical mood of passivity, which derived from an excessive sense of divine control of life and from the ecstatic experience of the Sufi saints, and he brought Muslim thought and social attitudes into contact with the modern world of progress, science, and technology. In his poetry, he extols ideals of human freedom along with a devotion to God. He denounces the excessive traditionalism of the Muslim community and calls upon people to dedicate themselves to a study of the new sciences that are now altering life so profoundly.

A third spiritual personality, this time from the Buddhist world, is Daisetz Suzuki (1870–1965). Suzuki brought Zen Buddhism into the twentieth century with such force that Zen is now considered a major element in contemporary thought, art, literature, spirituality, and religion. Zen is wonderfully alive in these times. The purists who point out the differences between ancient and contemporary Zen are quite correct, but what is considered a corruption from one point of view may well be a creative adaptation from another. The Zen of Suzuki is creative both in its fidelity to the tradition and in its presentation to the modern world. He has made it available as a way of realizing the unnamed possibilities within the human being. In a special way, Zen enables human beings to respond to the deepest spontaneities within themselves, the deepest intuitions of which they are capable. This influences the total range of human activities and has a special compatibility with our modern, scientific world, because it enables the mind to respond with new clarity to transrational scientific insights as well as to poetic insights. Zen also fosters an awareness of the ultimate mystery of things in the world and brings us into an intimate experience of the human, cosmic, and divine orders.

A fourth person who has shown how the spiritual meaning of the higher human process is carried on by science and our modern world is the French Jesuit and paleontologist Pierre Teilhard de Chardin (1881–1955). Among modern Christian writers he is outstanding for bringing together the traditional spiritual and modern scientific worlds. He is the first to penetrate fully into the sacral aspect of developmental time of the entire evolutionary process—both of the universe and of the Earth. He is a modern of the moderns, a traditionalist of the traditionalists. His deepest inspiration is undoubtedly found in the epistles of Paul, especially in the notion of the cosmic Christ pervading the universe. This early spiritual vision has shone with new brilliance in the sublime interpretation Teilhard gives to the discovery of evolution in the modern period.

From these extraordinary personalities we can easily perceive that there is both continuity and discontinuity in the movement of religion from the past into the present. This must be understood as creative newness emerging out of life-giving traditions whose vitality is by no means exhausted but which expresses itself with new vigor as humanity unfolds its full reality through the centuries. Religious possibilities expand as human life expands into new dimensions. The one requirement is that religious personalities now commit themselves to understanding and transforming this new world with the same vigor with which their predecessors understood and spiritually transformed the world of their own times. And now the challenge is how to do this within ecological limits.

CHAPTER 2

Religion in the Global Human Community

WE ARE presently creating a multiform global community as an effective and encompassing setting in which each person and each particular society finds a comprehensive context for existence. Within this global society of humankind, each person becomes heir to the fullness of past human cultural achievements, participant in the convergent cultures of the present, and, according to capacity, maker of the future. This convergence of the present, the consequence of scientific and technological improvements in travel and communication, has not so far been characterized by any dominant religious or spiritual motivation.

Yet it can be seen that exterior convergence does not necessarily bring either interior communion or cultural enrichment. An effective development preserving and enhancing the human quality of life requires a sensitivity to deeper forms of communication between subjects. For this reason, an understanding of human interior and religious life and the reconciliation of religious traditions with one another become matters of urgency.

In fact, a vast amount of literature is being produced concerning the religions of every corner of the world, notably from tribal Africa, South America, and the Pacific Islands, as well as from Asia and Europe.

However, this literature is still governed more by the methods of the positivist social sciences than by the humanist disciplines. Despite the urgency of the work and the benefits to be derived from understanding basic issues of divine-human presence, Western theologians, for example, have given little attention to interpretive problems emerging from the meeting of spiritual traditions. More generally, Western studies are not at present bringing about a desirable communion of peoples nor do they enable the various societies to provide one another with needed interior nourishment. What is being done to enable religious traditions to be present to one another in a truly human manner is often brought about by people who are seeking personal and emotional contact with diverse religions but who have little intellectual background.

To strengthen these associations and make them more effective, it seems desirable that those engaged in religious and spiritual studies be more creative in exploring and interpreting the larger cultural order. The documents with which they deal, the religious and spiritual events they observe, and the spiritual disciplines they describe so well all are associated with profound interior experiences that should flow forth as serious cultural influences within our society. Yet there is an amazing capacity of scholars to "defend themselves against the messages with which their documents are filled."[1]

Composed within the context of profound realism, intimate life involvement, and struggle with the most destructive aspects of the human condition, these messages function best and become most intelligible within that context rather than simply as academic discussion. This is the time, indeed, when the deepest meaning of these ancient traditions should emerge, for all humanist traditions presently seek renewal by intercultural and interreligious dialogue.

Since students of the contemporary social sciences are not generally comfortable with the humanist phase of their study, they tend to encompass the human order of things within empirical science. This is one way of seeing the present situation of anthropology, the social sciences generally, and even religious studies. These disciplines derive their method, their mood, and their objectives from the social sciences, which themselves derive from the objectivist methods of Newtonian physics. Within this context, the human quality of life as previously known is considerably diminished, the intuitive experience of the real stifled, and the will and capacity for grandeur lost. Humans lose their place at the center of the real and end up in ash cans on the stage of Samuel Beckett (1906–1989).

The question of the humanist and scientific dimensions of reality constitutes a crucial issue even as regards the physical survival of the human being. Scientists realize that their achievements, so grand in their scope and reach into outer space, have not enabled us to deal with the full magnitude of the human order, especially its interiority. The dominance of the social sciences powerfully influences religious studies today. Their vigor exhausted by analytical research, scholars of religion often find it difficult to push their studies onto the level of comprehensive humanist interpretation. Perhaps they are too close to the material for this deeper type of reflection; possibly the professional humanist, philosopher, or theologian is too far away from it.

Thus many of the great scriptural texts of the world, long available, remain without depth of interpretation or cultural influence. They seem not to have the vital impact seen in the movement of Buddhist texts from India to China, or Buddhist and Confucian texts into Japan, or the meeting of Christian Biblical texts with Patristic Hellenic thought, or in the Western accommodation of Christian tradition to Aristotelian treatises.

Thus in the present, insufficient thought is given to deeper levels of cultural creativity. One reason, perhaps, is that many Western scholars have no vital relationship with the deeper cultural currents of their own traditions. Their scholarly life is without relation to any experiential roots. Despite much excellent work in the translation and explication of diverse religious literature, we do not produce scholars of the Erasmus type, although this result is what we might eventually hope for. Creative personalities of such dimensions are arising more effectively out of scientific rather than religious studies. Not primarily due to deficiencies in the work of research scholars, this dearth of creative religious leadership is due, rather, to failures in understanding and communication of those supposedly in vital contact with spiritual and humanist currents of the West: philosophers, theologians, and cultural historians.

What is needed is an interaction of textual study and research with true humanistic insight and imagination. Present limitations in interpretation prevail in the imaginative and emotional orders rather than in technical skills of translation or the collection of research data. In the words of Mircea Eliade (1907–1986):

> It is not necessary to let oneself become paralyzed by the immensity of the task; it is necessary above all to renounce the easy excuse that not all the documents have been conveniently collected and interpreted. All the other humanist disciplines, to say nothing of the natural sciences, find themselves in an analogous situation. But no man of science has waited until all the facts were assembled before trying to understand the facts already known.[2]

Eliade's words are paralleled by similar statements concerning scientific insight by René Dubos (1901–1982): "Many great experimenters in all fields of science have described how their ideas were determined in large part by unanalytical, visionary perceptions. Likewise, history shows that most specific scientific theories have emerged and have been formulated gradually from crude intuitive sketches."[3]

Like all studies, whether scientific or humanistic, religious sciences need visionary perceptions, artistic awareness, cultural creativity, and the ability to respond to that depth of human consciousness "below the analytical level." The need is apparent especially at the present time, when global communities of scholars and a complex of cultures come together to establish the abiding context within which human life is to be lived and humanist studies carried on into the indefinite future.

Whatever the magnitude of the task and however vast the required imaginative range, we cannot really say that the work is proportionately so much greater or its objectives so different from the objectives sought and mission fulfilled during the early history of the classical civilizations. There was then, as now, the challenge of awakening to human meaning and purpose within a large and encompassing universe. The most primordial intuitions of humankind, as expressed in myths and spiritual disciplines, communicate to us across the ages—at least in outline—this cosmological context for cultural development.

In classical civilizations as well as in indigenous traditions, the comprehensive worlds of the divine, the cosmic, and the human were intimately present to one another. This, for example, is the very context in which early Chinese civilization emerged, as we see from the opening passages of the *Book of History*. Here is portrayed in extensive detail the human quest for integration of Heaven and the Earth, the effort to encompass the cosmos in its full extent and to order human existence in relation to natural systems.

There is, of course, much greater complexity in present cosmological, historical, and cultural processes than in those dreamed of by these ancient civilizations. Creation of a future world will always require new ways of integrating the human cultural complex in historical time and global extent. But it should help to know that humans have generally, from the earliest Neolithic period, functioned on a comprehensive cosmic plane, as evidenced by both archeological remains and living tribal societies such as Native Americans and the classical societies of India and China.

Here we observe that the work of ordering the world within its human context has, since its earliest period, been largely the function of hermeneutics, the most ancient form of human wisdom: interpretation. The great civilizations first read the text of the real from cosmic phenomena, then interpreted it, a reaction leading to the composition of scriptural texts handed down through the ages as revelation second only to the cosmos itself.

Ever since that time, humans have sustained and developed the greater cultures by constant reinterpretation of these ancient written scriptures in the light of new historical experiences. The constant renewal of civilizations, their very life process, has been associated with, and largely governed by, reinterpretation of these same texts. In our own present time, human beings must once again reinterpret their basic scriptures, this time in a context embracing them in multiform human traditions reaching from East to West on a global scale and in a historical time sequence witnessing the ongoing developments throughout the centuries and their present convergence. This present historical-cultural convergence must be seen as the primary context in which cultural traditions and religious studies now become maximally aware of themselves.

What must be sought for in the new hermeneutics is the recovery, through critical processes, of a second naiveté, an earlier interior experience of a harmonious and luminous universe, associated by the Chinese with the "lost mind of the child." The manner in which this perspective is achieved must, to a large extent, be through language. This is why language studies are of such great importance and why language and its most ancient literatures must be constantly reexplored to yield their full original luminous meaning. Today's religious scholar deals with the greater part of the world's most sacred literature in an era that is becoming the supreme scriptural age so far in human history.

The ultimate achievement of scriptural scholarship is the recovery of what Paul Ricoeur (1913–2005) designates as the fullness of language. We do not awaken to consciousness in a blank universe. We awaken in a universe wherein the cosmic script is already written, a universe in which the written scriptures have already been composed, a universe in which we discover, with Wang Yang-ming (1472–1529), a third scripture imprinted within our own being. Each of these scriptures—the cosmic, the written, and interior awareness—responds to the others, evokes the reality of the others, and is interpreted in their light. These three together guide us in our self-creation, our humanization. Out of these three scriptures the human cultures have been born and sustained and, when these cultures have declined, it is out of these three sources that they have been called back to life and renewed from century to century. Once more, in our day, all cultures face a challenge of severe proportions as the scientific assault on humanist and religious methodologies continues apace.

In the overall view, this confrontation is perhaps the special period of the third of the three scriptures, the inner awareness. This awareness, however, can come to life only through vital contact with the other two scriptures. Thus the period prior to the composition of early verbal and written scriptures is the period of unarticulated but deeply felt human response to the scripture of the cosmos. From it came the verbal scriptures so familiar to us over the last three millennia. Now we are in the third period, the period of inner awareness that reads the verbal scriptures within new scientific realizations of their context, in an evolving world and among diverse societies but without losing the older humanist and religious insights and values.

This is the crisis to be faced: the demythologizing and alienation from our humanist scriptural foundations, a period during which the scriptures survive only by objective, analytical study, more within an academic context than as part of the realities of human existence. This period of demythologizing is perhaps coming to an end at the same time as the period of cultural isolation—and at a time when the inherent limitations of technological achievements are being recognized.

At such a time, the interior awareness, this third scripture, awakens us to the need for renewed contact with the other two scriptural traditions, the written and the cosmic. A new period of scriptural vitality becomes possible. This is the meaning of creative movements emerging at the present time that may give this generation hope that a new humanism is being born. This reborn humanism will be the context in which religious studies will be carried out in the future. For the first time since its emergence in the historical process, Western culture has an opportunity to establish itself within a functional global complex of cultures in a spiritually cooperative attitude, rather than a spiritually antagonistic or competitive one.

The traditional antagonism has been largely the result of ignorance on the part of Western religious thinkers concerning the real nature and intent of the religious thought and practice of other peoples. But now, because of new types of religious studies, an extensive range of human religious development has been investigated in all its variety, from its simplest to its most complex forms. What emerges from these studies is that the various religious traditions, with all their differences, have much to say to each other. The beliefs and practices of each illumine the beliefs and practices of the others.

This sharing becomes evident in the case of Christian thought, which now has a broad range of data in world scriptural traditions upon which it can draw to deepen its own understanding of the divine and human orders and the relations between them. These data must now be incorporated into discussions of the divine reality in itself and in its relation to the phenomenal world. They must be drawn into discussions of the human condition, revelation, redemption, incarnation, the savior personality, faith, grace, sacrifice, rebirth, interior spiritual discipline, divine union, sacred community, communion with the cosmos, and final beatitude.

None of these can any longer be studied satisfactorily within a single tradition and without data from other religious cultures. With such data, Christians find not only that their own traditions can be further identified in their distinctive characteristics but also that their understanding of their own traditions can be considerably broadened, perhaps even more than biblical understanding was clarified and extended by contact with Hellenic thought.

Yet all of these advances require the development of a new and more satisfactory hermeneutics, the area of greatest urgency for scholars of religion. Extensive research remains to be done on materials not yet adequately interpreted. This advance can come about only by an increase in scholars with adequate cultural-historical backgrounds and the intellectual insight and imagination to incorporate existing data into more meaningful contexts.

Students of religion who work from a philosophical or theological basis are generally not acquainted with actual religious beliefs and practices on a sufficiently broad scale, nor do they always have the type of interpretative skills necessary. On the other hand, those involved in scientific research frequently are unable to go beyond their data and consider that "subjective" appropriation and interpretation of data contaminate pure scientific knowledge. This transition from scientific knowledge to subjective human meaning is indeed a big step and needs to be managed with extreme care. Yet any study of the human should be recognized from the beginning as a subjective activity involving the knowing person, the reality known, the means of knowing, and its purpose. The "scientific" process in the acquiring and organizing of data enables us to become more clearly aware of the human mode of being, but only when the construction of scientific methods and categories do not diminish or minimize the basic human quality of the process.

If, as regards studies of the religious life and literature of one people by another, subjective communication is not taking place, if religious personalities cannot speak to one another in terms mutually helpful in managing the human condition, if traditions cannot clarify for one another what it is to be human and assist one another in carrying out a redemptive or liberating transformation of the human subject, if the sacred space of the one is impenetrable to the other, then, it would seem, the study of our human religious and cultural traditions comes to an abortive conclusion. Yet we must admit that we have not yet satisfactorily learned the art of interreligious communication.

While history indicates extensive interreligious and intercultural conflicts and tensions in the past, it also presents extensive examples of religious and cultural influences successfully passed from one society to another. Indeed, intercommunion of traditions has taken place so widely as to seem almost universal. It would be difficult to identify any spiritual or cultural tradition as a "pure" tradition, an absolute self-creation. What has not yet been done and what is much needed is an understanding of how this religious-cultural process can be carried out more effectively in the conditions of modern times. As models from the past, we have such great periods of spiritual communication as the Patristic era of Christian development in the Mediterranean region, the assimilation of Buddhism into the Chinese world, and the spiritual and cultural interaction in the development and spread of Islam across south and central Asia.

The difficulty is that now the situation has grown extremely complex, with the impingement of all traditions on one anther in a flood tide of religious data and influence, on the one hand, and a lack of authentic human and spiritual communication, on the other. There is no proportion between the amount of data and the quality of genuine religious intercommunication taking place. The disproportion is due to secularist indifference, vigorous fundamentalism, and professional neglect, especially by Christian theologians.

Still, the entire dynamics of the modern world is, to an increasing degree, throwing people and societies together in close exterior proximity but without the capacity for interior communication. This failure to communicate interior human realities is the main difficulty we face. Formerly, peoples and traditions met on a limited scale, with limited personal contact, but within a fundamentally human and religious order of life. Now, communication has been intensified on the exterior without satisfactory deepening within: thus the reduction of human relations to economic, political, and social orders, with an overlay of the aesthetic and the spiritual. It may seem rather distant to speak about interreligious hermeneutics in such a context, but certainly we will need to interpret our deeper selves to one another if the human venture is to meet the challenges of the future for cultural and ecological survival.

Much has, of course, already been accomplished in the spiritual order. If we look to the larger cultural context of the present, we can observe a powerful gravitation of traditions toward one another: thus the popularity in the West of a variety of religious cults and spiritual disciplines from various parts of the world. There is a return to religious symbolism, participation in rituals, engagement in difficult and demanding meditation exercises, quests for the guidance of spiritual masters, and extensive reading in various world scriptures.

However, these activities are mostly a matter of personal interest and are unrelated to traditional religious establishments or to official teaching. In America, at least, we seem to have few theologians of recognized competence in world religions, although among Asian scholars and cultural anthropologists there is extensive commentary. Western theologians seem to concern themselves entirely with the challenge of secularism and prevailing philosophical currents of more recent Western derivation. Thus participants in a variety of exotic spiritual traditions are often without proper guidance. Still, even if they do not lead to substantial religious or cultural development, these activities at least bear witness to basic spiritual attractions of various traditions for one another and offer some indication of new forms of spiritual development that may someday take root in the West.

If there is, on the part of some, spiritual immobility and resistance to such sharing in the face of the meeting of peoples, there is also a trend toward intimate participation. The difficulty of the one is that spiritual traditions are not developing according to the historical dimensions required in this exciting period of human development and thus are not contributing significantly to the needed communion of peoples and traditions. The difficulty of intimate participation is that it makes too little distinction in responding to religious phenomena and thus finds no sustaining life program, nothing that can be elevated into a movement toward the cultural renewal demanded at this moment in history.

Here is the task of hermeneutics, to interpret traditions to each other so that, while keeping each tradition distinct, it brings them into the creative presence of the other. Only on the condition that each remains inviolably itself can the traditions help one another. Falsification or overly facile identification in an undifferentiated religious or spiritual context can cause the movement of interpretation to lose its effectiveness. Both continuity and discontinuity must be preserved. Diversity of traditions must be renewed while common spiritual space is established.

Each tradition will have a particular, limited microphase and a macrophase open to the broader meaning or resonance of comprehensive human traditions within which it now begins to function. This distinction of microphase and macrophase might very well serve to differentiate the larger, more universal phase of a tradition and its particular, limited phase. So it is with Christianity. It has its specific institutional phase, its community of the baptized; yet beyond these aspects there is the larger community of humanity with which it is associated, not simply in a purely spiritual manner but by way of observable Christian presence.

And so it is with all the religious traditions. Yoga, for example, has its microphase as expressed in association with *Samkhya* philosophical dualism and its macrophase in relation to spiritual disciplines associated with many of the world's religions. We might speak, also, of the microphase and the macrophase of key terms used in various traditions: *brahman, maya, nirvana, karma, dharma, li, tao, t'ien, jen.* All have their origin and primary significance within a certain historical and religious context yet are now being universalized to enrich the religious and spiritual vocabulary of the global human community.

For example, we have already a rich variety of terms from a complex of traditions. Each reveals something not quite the same as that denoted by other terms, indicating that the experience itself is not quite the same. Yet the variety of terms from such a complex of traditions significantly assists us in our efforts toward identification of ultimate reality, indicating a "final term of reference" in the order of the real or in the order of human consciousness.

Proper use of these terms requires a special skill—a skill we are only now acquiring. It must also be added that when such a complex of terms is brought together, they all become changed and begin to enunciate things never enunciated before. The linguistic or theological "purist" is prone to object to such "abuse" of sacred words. Yet neither history nor culture nor language knows the type of permanence or constancy sought by those who would isolate language or thought from the temporal conditioning to which all things are subject. We cannot accept the position that languages are ultimately opaque to one another nor the attitude that thought systems or religious traditions or spiritual disciplines are incommunicable to one another.

Precisely here is where the creative intuition or visionary perception begins to assert itself. The horizon is open, extensive linguistic materials are available, and a larger vision begins to emerge within the global community. Past traditions have each made important contributions; the present has brought about a convergence of spiritual currents and their modes of expression. Now is the creative moment, the moment on a global scale such as that time within Western civilization when Paul began to write his epistles to the people of the Mediterranean world, or when Augustine enunciated his vision of a Christian historical order, or when Thomas brought Aristotelian terminology into Christian theology so that it could say things Aristotle himself never dreamed of saying.

There are no laws for such moments; there is no way of indicating the direction religious insight should take. These choices are determined by persons of superior religious imagination who understand the complex of past traditions and the opportunities of the present and who can construct a comprehensive religious vision adequate to present historical circumstances. Such persons choose and shape language suited to this purpose. Subsequent generations of scholars will fill out details within the established context until another period of major transformation arrives.

Lacking such a resolution of present problems of religious hermeneutics and religious thinkers of this magnitude, those in religious studies can only proceed within the limits of their ability toward creation of a truly functional religious vision. If we indicate, then, the need of Christian thinkers, as a first step, to incorporate data of other religious traditions into their study, then the second step, the development of a new science of hermeneutics, can be taken.

A third possible development concerns the need for studying Christianity according to the norms and methods of the history of religions. In recent centuries, the Christian tradition has been dominated by biblical expressions or their theological expositions from the medieval period. As the basis for explaining and defending beliefs considered essential to a Christian life orientation, this theological tradition achieved remarkable success, immensely sustaining and developing Christian thought and culture. At the same time, it severely limited Christian expression and the range of religious experience.

The only extensive cultural, religious, or philosophical influence on this Christian theological tradition so far has been that of Hellenistic philosophy from Platonic, Neoplatonic, and Aristotelian sources. More recently, there have been influences from sources such as phenomenology, the existentialism of Martin Heidegger (1889–1976), and, lately, the process thought of Alfred North Whitehead (1861–1947). Thus one of the most effective ways at the present time of invigorating

Christian thought might be to explore Christian tradition from the wider anthropological and cultural perspectives of the history of religions. From this perspective, the major outlines of biblical and Christian tradition emerge in a much clearer light, the terms and norms of comparison having been previously far too limited. Such a view of Christianity has already emerged in general studies of the history of religions and its companion discipline, the phenomenology of religions. Yet these early observations have not gone beyond basic observations; there is a need for a more thorough historical and phenomenological study.

A fourth avenue of growth and development for Christianity is increased study of the nature and limitation of symbolism, especially archetypal symbolism as set forth extensively by Carl Gustav Jung (1875–1961) and others. Religious life in recent decades is recovering from a devastating period of rationality and scientific analysis. The limitations of the strictly rational processes of the mind are, however, once again recognized, along with the inability of rational demonstration to evoke emotional response or sustained application of human energies. The more profound aspects of human awareness and feeling come from a depth of our being to which we ourselves have but limited access. Only from symbols emerging in our dreams or arising spontaneously in our consciousness do we know this depth. Cultivation of this inner capacity for understanding and responding to intuitions of the mind constitutes the basis of creative genius, whether in intellectual or aesthetic human life. It is especially important in spiritual and religious life, which, in its essential qualities, involves communication not only with transrational but with transhuman modes of the real.

Both from the standpoint of understanding and of efficacy in the spiritual, social, and historical orders, the recovery of contact with imaginative processes is of supreme importance. Also, in the renewal of the religious and humanistic phases of contemporary life and in the evocation of a greater intercommunion of traditions, hardly anything is more helpful than this new appreciation of the imaginative, symbolic, and mythical.

Of the various scriptures here mentioned, the cosmic, the written, and interior awareness, we might say that all come together in the common world of symbols. Rebirth symbolism, impressed upon us so powerfully by the sequence of the seasons and the periodic death and rebirth of living things, has entered into the myths, rituals, and sacred literature of peoples so extensively that it must express a common interior awareness. The specific manifestation and meaning of rebirth symbolism differs from one people to another, but the consciousness that, to achieve human status, we must undergo a death-rebirth experience has far-reaching implications.

Another symbol important in the spiritual intercommunion of peoples throughout the world is that of mythic narrative, in which human existence is seen as a story or drama depicting a journey wherein the human personality encounters obstacles to be overcome, demonic powers to be thwarted, and even death to be endured before, with the aid of superhuman powers, the true self, the divine presence, the grail, or the jewel of great price, everlasting life, is attained.

This journey motif dominates the ancient Babylonian epic *Gilgamesh*, the Homeric epics, and the Hebrew Exodus legend. It is found in the wanderings of Buddha, in the fictional account of Hsüan Tsang's journey to India, and in the Hopi Indians' tracing of their origin and destiny. Its Christian mode appears in Augustine's (354–430) story of the rise, growth, and termination of *The City of God* and in the phases of Dante Alighieri's (1265–1321) *Commedia*. In all of these accounts, external pilgrimage is the symbol of the journey of the individual soul, the particular society, and the human order into those interior depths wherein the sacred presence shines forth, all peril is surmounted, and final security is attained.

A fifth area of study deserving consideration is the religious experience known in biblical and Christian terms as *revelation*. At least so far as the West is concerned, a more comprehensive study of divine revelation to humans remains an important need for broadening religious perspective. Understanding the full range and depth of revelation as a religious event will require both intensive research and greater insight into resultant data. These will provide the basis for a collection of world scriptures out of which new thinking of the religious order can be done. If, so far, only a few efforts have been made in this direction, we may hope that much more will be done in the future.

At first glance, revelation appears to be primarily a Christian concern and a human-divine communication so intimate as to be indescribable at the intellectual level. Revelation is, nevertheless, a phenomenon of universal importance and one of the most fruitful points of religious discussion, at least for theistic religions. Within the protocols of revelation, the various scriptures will not equate in any univocal sense. The revelatory nature of Hindu or Buddhist scriptures will differ in form from that attributed to Jewish and Christian scriptures, but the basic concepts cannot be totally unrelated. This kinship, as research identifies it, will provide the basis on which future thought must build and mutual validation of various scriptural traditions considered.

Finally, diverse religious personalities communicating major religious or spiritual teachings to a people might also be given mutual recognition. These include contemporary figures and earlier thinkers such as Aurobindo, Iqbal, Suzuki, and Teilhard, discussed in the first essay.

What can be seen from these suggestions is that, after the weakening of spiritual and cultural consensus in recent times, the former isolated situation can hardly be reestablished. The various religious traditions are irrevocably altered in their individual and collective significance. Our entire spiritual situation—the very mode of our religious consciousness—is changed. Suggestions that the sources of revelation be broadened for Christian theology, that multicultural and multireligious hermeneutics be now a central question, that the Christian tradition be studied according to the norms of the history of religions, that comprehensive study be given to divine revelation, and that a world scriptural collection be developed—these might be among the most significant issues to occupy our efforts in the immediate future. These will shift the content of Western Christian awareness into a new context, the multiform global religious tradition of humankind.

If, formerly, Christianity was studied from the inside out, it should now be studied from the outside in. In this situation, historians of religions are among the foremost religious creators of the twentieth century, primarily responsible for evoking consciousness of a universal human religious heritage in all its diversity. Also, they are responsible for creating the conditions in which extensive interaction of religious traditions and cultures is begun, to be continued indefinitely into the future. Of all the forces at work in the modern world, it is doubtful if any is more powerful than an enlightened religion in awakening ancient traditions to new phases of development. For none of the traditions is in itself complete. Seeming for long periods to remain unchanged, these traditions are now awakened to development and renewal as seldom before in their history. They begin to realize that they are not entirely stable forms of life but rather developmental processes that have changed considerably in the past and are destined, perhaps, for even greater change in the future.

Yet since these traditions have all undergone extensive interaction with other traditions over the past several thousand years, we should not exaggerate the newness of modifications brought about in the present. The sense of novelty results only from a more heightened intercommunion of traditions due to the definitive nature of present-day intercommunion, its comprehensiveness, and its pervasive nature, all results of modern means of communication. These means are bringing peoples and traditions of the world into one another's presence to a degree never before possible.

Strangely enough, the very forces moving various cultures and religions out of the traditional into the modern world are exactly the forces enabling each tradition to recover contact with its most pristine forms and ancient literatures. Thus these traditions live more deeply in the past even as they move into the present and future. Each tradition is made more complete within itself, more integral with its primordial moments. From this earliest period, the historical movement of religions and cultures has been converging toward multiform global expressions in which each finds its place and each is in some manner present to the entire human society.

CHAPTER 3

Alienation

ALIENATION IS, in some sense, the oldest and most universal human experience. It is our human condition: the difficulty of discovering our personal identity and our proper place in the universe. Particularly in Western civilization in the nineteenth and twentieth centuries, humans have experienced the challenge of authentic existence while moving through a series of rapid historical transformations. Alienation of the workman from the means and benefits of his production was the central social issue from the 1848 manifesto of Karl Marx (1818–1883) until the dissolution of the Soviet regime in 1991. Alienation of the "authentic self" from the "false self" that we adopt has been a central issue in psychotherapy throughout the twentieth century. Alienation was especially severe in the countercultural movements of the radical left in America in the 1960s, both its dramatic protests against existing social structures and the "flower people" of that period, with their romantic-mystical rebellion against the harshness of industrial life.

In the opening years of the twenty-first century, we are experiencing a new alienation in our inability to relate effectively to the integral functioning of the Earth. This alienation, which results from an extreme anthropocentrism and dedication to consumerism, is causing the exploitation and devastation of the planet, supposedly for human advantage. Until recently, few people have realized the extent to which human fulfillment depends on the integral functioning of the Earth in all the grandeur of its natural landscapes—the forests, mountains, woodlands, rivers, and lakes—and in the wonder of its wildlife: animals, insects, fish, and songbirds.

Alienation from the world of nature has led us to extravagant expectations concerning the benefits of our modern technologies. These expectations have blinded us to the evils inherent in the very solutions to life's difficulties we were proposing. By using chemical fertilizers, we increased our grain harvest but destroyed the natural fertility of the soil. Because of our clearcut assault on the woodlands, the forests can no longer renew themselves. In relentlessly pursuing marine life, we depleted the abundance that was there for millennia.

In the last two centuries, as we have become more proficient in manipulating the nonhuman components of the Earth community, we have become progressively alienated from the most elementary awareness of our role and place in that community. We expected the entire universe to respond to us, the human component, as the ultimate reference and arbiter of value. Frustrated when we realize that we do not have control over the world around us, we sink into a deepening cultural impasse.

In becoming a commerce-dependent consumer society, we have ignored the essential elements and ideals necessary to sustain any viable human community. For example, by enclosing ourselves in automobiles, we have isolated people from one another and destroyed a certain sense of community. Moreover, we find that the distance between the affluent and the less well-off and from the impoverished is constantly increasing. We are isolated and alienated, both as individuals and as communities. We are held together mainly by the political-legal binding of the modern nationalist state and by our dependence on an industrial, commercial, consumer society.

In an earlier age, recognition of the primordial alienation of the human condition led to the development of mythic structures by which humans could make sense of their situation, mitigate its destructive aspects, and turn it into a liberating and salvific experience. Creation stories indicated the manner in which things came into being. These included paradise myths of the original human condition; myths of our fall from paradise, which resulted in alienation from ourselves and disordered relations with the larger order of things; myths of a divine or heroic slaying of the dragon; myths of human redemption through the deeds of the gods; and myths of spiritual rebirth and the reattainment of paradise.

Indra slew Vitra. Venus watched over Aeneas. Amida Buddha stretched forth his hand as a protective shield over all humankind. In these myths, humans possessed a way of dealing with the world of mystery, that world filled with forces both benevolent and terrifying and beyond rational understanding or human control.

Although physical deprivation remains a serious affliction even for our prosperous modern civilization, one of the most devastating aspects of the present is the loss of traditional symbols. In the Chinese civilization, there was the *yin* and the *yang* symbol of opposites. The dragon was a feminine figure symbolic of the primordial power of the universe. There was also *Kuan Yin*, the Buddhist goddess of mercy. These symbols provided support for a meaningful existence and the attraction that held things together. In India, the lotus was the all-pervasive physical symbol of the soul emerging out of the dark world of matter and of disintegration into the purity and exalted status of the spirit world. There was also the figure of Krishna, the divine being incarnated in human form to be present with humans. In the love affairs of Krishna, the Hindu found the symbol of the high mystical union of the soul with the divine. The river Ganges was the symbol of the final purification of the soul whereby the deceased entered into a transearthly future.

While humans were aware of their limitations, they instinctively understood that they were not isolated or alienated in any absolute sense. Rather, they understood that the deepest forces of the universe, both personal and impersonal, provided a method of reunification with their deepest self and a way to participate in the deeper natural and divine harmonies that constituted the full meaning of existence.

Spiritual rebirth ceremonies healed the experience of personal alienation and brought the individual or group to an integrated, even a divine, mode of being. The universal insistence among the various traditions that we must undergo a spiritual rebirth before we can be considered complete in our human reality is an indication that we require a transformation that goes to the very foundation of our being. This symbolic death and rebirth can be seen as a basic requirement for self-identification, for participation in the social order, and for entry into the deepest mysteries of cosmic-divine reality. Here the entire ritual process is ordained toward our authentication in the human modality of our being.

The Divine Mother symbolism so widespread throughout the world (especially in the Near Eastern civilizations and in India) helped cultures deal with the immense universe and the sense of personal isolation. In China, the Tao was seen more as Divine Mother than as Ruling Father. In East Asia, *Kuan Yin*, originally a masculine savior divinity, was gradually altered into a feminine figure, to provide for those deeper needs that humans feel for a compassionate heavenly personality bending down over the world of human affliction. This awareness of feminine affection and concern as a supreme revelation of being came upon the world with great power and removed the destructive experience humans had of themselves and of the universe in which they lived. Such creative, redemptive, and bliss-bestowing deities dwelt in great shrines before which humans could stand and feel themselves in the presence of that which fulfills their most fundamental needs and their most sublime spiritual longings. Humans participated in a divine order of things, felt its healing pour over them, and realized that they were much more than insignificant creatures tossed about in an uncertain and meaningless universe.

Yet, great as the myths, symbols, rituals, and revelations of this earlier period were, they were all finally inadequate to the task of bringing about a complete healing of both inner and outer alienation, unless they were associated with an interior spiritual discipline that led to an experience of the divine beyond all symbolic utterance or ritual enactment. This movement to an experience of divine reality within our own being brought about a new age in human spiritual formation. In a particular manner, it healed the last trace of the alienation that humans had experienced within their own being. One of the remarkable things about the classical civilizations of both Europe and Asia is the high destiny that humans set for themselves and that they considered the normal end toward which human life was directed: identity with Brahman, the experience of Nirvana, the living presence of Tao.

Such interior spiritual developments were widespread throughout the Eurasian world in the middle of the first millennium bc. The most powerful manifestations were in India and China, where a series of interior spiritual movements took place for which there is no real parallel in the West except, perhaps, in the ascetic efforts of the very early desert fathers. But this movement in the West lasted only a short time before the great monastic orders were formed, which, in the West, became centers of ritual observance as well as centers of interiority and contemplative disciplines. The Neoplatonic traditions of Dionysius the Areopagite enabled this transrational intuitive experience to survive in Western consciousness. But the special meditative techniques of interiority so highly developed in India were less evident.

These interior disciplines in Asia are best exemplified by the Yogic tradition, which concerned itself directly with authentic human existence. Yoga realized the depth of alienation that exists in the very structure of the human's own being. An authentic state of being required a release from the bonds of attachment to the phenomenal world and the attainment of a perfect *inseity*, a perfect indwelling within oneself. As is indicated in the Yoga Sutras of Patanjali, humans must attain to a state in which their interior spirit "is established in its own proper form."[1] This yogic *inseity* is the most dramatic effort of the Indian world to remove humans' alienation from themselves and to achieve authentic existence through the experience of total self-identity. This indwelling is attained not by improving the human condition but by a deconditioning process wherein humans are liberated from the limiting structures of their own being into that limitless freedom known as *kaivalya*.

Another effort at healing human self-alienation is found in the Vedantic Hindu tradition, wherein, again, humans find their true identity beyond the reality of their own limited being in that interior self which is the *Urgrund* of their existence. In the Vedantic literature, the supreme difficulty of the human is lack of self-awareness. A change must be brought about whereby the consciousness of an outer world and a phenomenal self is accommodated to a consciousness of the interior sense of the self, the hidden reality of the human. Only in this state of total subjectivity are humans able to overcome their false existence in the ephemeral world of *samsara*, the world of unending coming-to-be and passing-away, in favor of an authentic existence in the world of the absolute. The story is told of a lion cub lost from his mother and raised with a group of lambs. Feeling himself a lamb, he went about gamboling and bleating in the soft and gentle ways of his companions until, one day, he happened to get a taste of blood. At that moment he awakened to his authentic existence and let out a lion's roar, for the first time overcoming his hidden alienation from his true self. Thus the problem of identity is fundamentally the problem of interior awareness. The lamb is the phenomenal Ego, the lion the absolute Self.

In China, the Taoists sought to create an interior consciousness of the omnipresent and all-pervading Tao. Humans must present themselves in a condition of total passivity to the immanent activity of Tao. For while the absolute principle in the yogic world of India is in itself completely quiescent, the absolute principle of the Taoist world is the dynamic of reality and is attained by an attitude of human receptivity described as nonactive, as *wu wei*. Nonbeing and nonaction are exalted as the conditions by which humans attain their authentic reality and remove all self-alienation in the modality of their lives. Every activity of the human that bears the stamp of rational determination is an interference with the deeper spontaneities of the Tao. This applies even to the spiritual discipline itself, for this too can alienate the human if it goes beyond the development of an increasing sensitivity to the immediacy of Tao.

In China, Confucianism also had an interior spiritual discipline intended to help humans overcome their alienation. This alienation was seen as resulting primarily from humans' failure to cultivate their heavenly endowed nature. This nature bears in itself an inalienable integrity. Even if suppressed and distorted by the inner violence that humans impose on themselves, the authentic tendencies of nature remain. The difficulty is that humans become lost and do not know how to find themselves again. Mencius tells us: "Slight is the difference between humans and the brutes. The common man loses this distinguishing feature, while the noble person retains it."[2] Again he mentions: "Sad it is indeed when a person gives up the right road instead of following it and allows their heart-mind to stray without enough sense to go after it. When one's chickens and dogs stray, one has sense enough to go after them, but not when one's heart strays." He concludes that "the sole concern of learning is to go after this strayed heart-mind."[3] The Chinese were quite conscious of the experience that we describe as self-alienation. Recovery of the self was the fundamental object of human cultivation and of all spiritual discipline.

But while the Asian world was occupied with its interior disciplines, the biblical-Western world was seized by a mythology of eschatological expectation that looked for the final meaning of the human in a future event toward which all temporal processes were directed. This new mythology produced historical drives of amazing vitality. Within this context, the problem of alienation becomes a question of self-discovery within the time process. Since time is fragmentary and looks to a distant event as its goal of attraction, rather than to the wholeness of the present, the problem of self-discovery lies in acceptance of the fragmentary present as one in the sequence of historical stages through which human development must pass on its way to fulfillment.

This transitory nature of life demands constant adjustment to new human situations, new expressions of thought, and new emotional attitudes in humans themselves, as well as new structures in the social order and new ways of relating to the cosmic and divine orders. The increasing attraction of an eschatological point of completion brings about a constant acceleration of change as history moves forward toward its fulfillment. Such constant change produces a shattering effect on humans, who cannot adapt to new situations with the required speed. Even if we succeed, there is in the process a draining of life's basic satisfaction, its continuity, its security. There is a loss of present meaning, because each moment of time lies under the condemnation of the next moment. An iconoclastic attitude is developed toward the entire present order. Utopian expectations are intensified. When these are denied fulfillment, an unbearable frustration sets in. The vortex of change produces an explosive tension.

This is an inescapable result of a tradition founded on the dominance of our Western time experience of reality over that of an Eastern spatial experience of reality. We greatly need to slow down the sequence of time changes by increasing our spatial awareness. Space is complete; time is fragmentary. Space is contemplative; time is active. Space has a present center of rest, time a future goal of attraction. Space is serenity; time is anxiety. The world of space is the world of nature, with its rhythms of rising and falling, its seasonal renewals, which can be easily ritualized and coordinated with the phases of interior renewal in the human. The world of time is the world of history, wherein humans pass through a succession of death-rebirth experiences in the form of historical convulsions as they advance on their way to the final eschatological convulsion, which introduces the fulfillment of the redemptive historical process.

We are at present in one of these convulsive moments in the unfolding of the human developmental process. Like the "flower people" of the 1960s, we might wish for the quiet of an interior transcendental experience by which we could rid ourselves of the feverish strife of external processes. Yet we must accept the fact that the temporal process has become the dominant human process and must be dealt with. In a world that we experience as both a coming together and a moving forward, we need a new planetary symbolism: a time-space, active-contemplative, prophetic-mystical, future-present, human-divine symbolism that will provide a context for discovering the journey toward self-authentication that humans deeply desire. To some extent, this is already available in the death-rebirth symbolism that is one of the most universal ways by which humans adapt to the reality of their own being and find their proper relation to the larger world in which they live. It is especially valuable in reading the human historical experience at the present time.

If it is to fulfill its historical function, this present world, so far distinguished by its science and technology, must now become distinguished by its spiritual depth and the richness of its human qualities. It must in some manner make present the paradisal aspects of the eschatological community. At its final moment of completion, the terminal process finds its fulfillment in the wholeness, the peace, and the paradisal qualities associated with spatial rather than with temporal experience. The main function of contemporary spirituality is to create this interior paradisal space, in which we can breathe the refreshing air of eternity and thus save ourselves from suffocation in time. A foretaste of such an experience is what we seek in our vacations to the shores of expansive oceans or to distant mountains. The future must be felt as already present. Humans cannot long sustain alienation from the ecstatic bliss that is the culmination of all great cultural traditions. If this is not granted in the immediacy of the present in its legitimate reality, then we will seek illusory fulfillment in whatever ways are available to us. We must live in paradise not tomorrow but today.

Thus the traditional and widespread symbols of the creative, redemptive, and paradisal experiences must be restored to validity in interpreting the present human situation. The primordial symbolism and the spiritual disciplines of the ancient civilizations were not flimsy and ephemeral productions of passing human fancy. They are tough and enduring realities capable of carrying the weight of the centuries and the larger hopes and destinies of humankind. They are more real and more needed today than ever before, if human life is to have the serenity and vigor required to move the peoples and civilizations of the world into deeper integration with Earth processes.

Yet these religious traditions cannot do what needs to be done solely with their own resources. The integrity of the life systems of the planet, diminished in the spread of humans across the planet, must now be restored. Neither the historical human nor the distant divine nor both together can provide the deeper integration we seek at present. This integration cannot be attained simply by the human, for indeed the human is not the ultimate measure of its own fulfillment. Neither can this integration be attained through the divine, because the divine cannot be its own manifestation. For this a phenomenal world is needed. A universe. An integral self-organizing universe.

Integral self-functioning is a dimension of the universe from the beginning. In the phenomenal order, the universe functions as the supreme reality. The universe is the primary sacred community. In the phenomenal order, it is the universe that is self-referent as regards reality and value. Every other mode of being is universe referent.

It takes a universe to bring humans into being, a universe to educate humans, a universe to fulfill the human mode of being. More immediately, it takes a solar system and a planet Earth to shape, educate, and fulfill the human. The difficulty in recent times is that the concern of the human in all the various traditions, with few exceptions, has been focused almost exclusively on interhuman and divine-human relations. Human-Earth relations have not been given the comprehensive consideration needed. That is where our contemporary challenge is located.

There is a special urgency just now because of the power of the empirical mode of knowing that has been developed by the scientific effort of Western civilization. This capacity for understanding is based on empirical observation interpreted through rational insight combined with exceptional powers of imagination. The scientific community has developed this powerful mode of understanding but in many respects has exacerbated the situation: first, by its explanation of the universe simply through mechanistic forces acting in a random fashion; second, by its subservience to the exploitation of the Earth through its alliance with technological invention and commercial exploitation.

Because we have relied on empirical ways of knowing for so long, we have become alienated from the immediacy of the natural world. We thought primarily of using our new knowledge for our own human advantage and with an arrogant disregard for the well-being of the planet. Although in some ultimate sense we should still concern ourselves with human-divine and interhuman relationships, we now urgently need human-Earth integration. That is where the alienation is at the present time. We are killing many of the more elaborate life systems of the Earth. The divine-human relation has attained magnificent expression throughout the past. Its need has been sufficiently authenticated. Although the interhuman relationship is far from fulfilled, it has received significant attention through social-justice issues. The difficulty at present is with human-Earth relations; the two remain alienated from each other.

This difficulty with human-Earth integration is intensified because scientists have insisted that their inquiry, based simply on measured modes of explanation, leads to a materialist universe that came into being by purely random processes. While scientific understanding of the emergent processes of the universe and its modes of functioning have been outlined by the scientists, they have been unable to identify the human role in that process. Until recently, the human was simply an observer disconnected from the process being observed. Now with the emergence of quantum and relativity theories, the place of intelligence in the total process has become a subject of discussion, especially in light of the cosmological anthropic principle, which indicates that the human was an intended outcome of evolutionary processes. What is even more significant is the extent to which scientists have become subservient to the economic exploitation of the entire planetary process, with little recognition of the intimate dependence of human spiritual and religious experience or the imaginative, emotional, and intellectual life of the human on the integral geological and biological functioning of the planet.

A WAY FORWARD

In our modern empirical inquiry into the origin and structure of the universe, we have attained a new understanding of the universe as the ultimate referent of every mode of being. Indeed, science loses its validity if this bonded relationship of every mode of being in the universe to every other mode of being is not accepted. We can explain nothing if we cannot explain the whole. Our explanation of any part of the universe is integral to our understanding of the universe itself.

It is most significant for the human to experience itself as brought into being, sustained in being, and fulfilled through the comprehensive universe. This coming forth from a physical and nurturing source requires also our return to the universe as our final destiny. In more proximate terms, whatever is said of the universe applies proportionately to the Earth context. The various components of the Earth form a single integral community.

In this manner, our alienation is definitively overcome within a unified understanding of ourselves, of the universe, and all the forces present therein. The genius of our times is to join the physical identification, experience, and understanding of the Earth, given by scientific inquiry, with the traditional mythic symbols and rituals associated with the Great Mother. To appreciate both of these in their proper relationship is to overcome our alienation from the universe and from the Earth. This understanding should evoke the emotional and imaginative sympathies needed for the sensitive care humans need to give to the natural world and also provide for our aesthetic excitement and celebration of the natural world.

What is needed is to bring back the ancient symbol of the universe as the Great Mother. This symbol has at times been interpreted more at the level of the planet Earth than in the more comprehensive sense of the universe. Yet its full expression is found in the sense of the universe as the ultimate referent in the phenomenal order of existence. Only the universe is self-referent. This is the proper role of the Great Mother: to be the primordial source whence the vast diversity of beings in the universe comes into existence.

Similarly, the death-rebirth symbolism is especially significant at the present time, as we are witnessing a death phase of the planet. The Earth that blossomed in such grandeur during the late Cenozoic Era has in the twentieth century experienced devastation that is extinguishing immense numbers of Earth's most sublime creations. Death on such a scale, caused basically by a single species with rational understanding, has, at least in its modality, never before been experienced over the total course of Earth history. For this death in its present form was brought about by human activity, even though we were not always aware of the effects of our actions.

So severe is this devastation of the planet that we can conclude that survival of complex ecosystems in any integral manner requires that a type of rebirth occur, a rebirth that could be identified as the emerging Ecozoic Era. Even the extinctions at the end of the Paleozoic Era some 220 million years ago and the extinctions at the end of the Mesozoic some sixty-five million years ago were different from and less devastating for the more evolved forms of life than the extinctions of today.

Given the immense numbers affected and the severity of their nature, to interpret events at this order of magnitude we need the traditional death-rebirth symbols. Yet the death-rebirth that we are here concerned with brings us back to the original creation of the universe and the powers present at that time. The seasonal sequence of death-rebirth within nature's cycles of renewal is itself being terminated. We are into what has been referred to as "the death of birth." Human affairs until now have been thought of within this seasonal cycle. The cycle was never in question. The only problem was the insertion of human activity into the cycle. Now the cycle itself comes into question.

We little realized what we were doing when we began to interfere in the processes of the natural world. Through our technological skills and our capacity for genetic engineering we have sought authority over "life" and "death" themselves. Now the extinction of life as this has found expression over the last sixty-five million years of the geobiological period, designated as the Cenozoic Era, presents us with an issue far beyond anything that we can manage in any effective manner.

Yet we must think about and respond to the urgency of a renewal of the integral community of life systems throughout the Earth. Renewal is a community project. Only the community survives; nothing survives as an individual. Here our sciences reach their limits, both in understanding and in efficacy. The shallow regions of control that humans possess over the natural life systems are vast in their powers of devastation but pathetically limited in their powers of renewal. We can take away, but we cannot give life. We can only accept, defend, foster, and occasionally assist in healing the world of the living. The skill to carry out these four functions will be the great skill of the future. Our hope is that the vital forces within the diminished life systems of the Earth can themselves renew, at least in part, the creativity shown over the centuries. For example, we see the system's creativity at Krakatoa, the island between Sumatra and Java whose life systems were almost completely destroyed by a volcanic explosion in 1883 and then with marvelous recuperative powers returned to life. This resurgence occurred independently of any human assistance.

To bring about renewal on the larger continents, where human presence and the industrial process have exhausted the land and fouled the air, polluted the rivers and ruined the forests, and to restore the great shoals of fish that once filled the seas: this is the challenge. We need to understand death-renewal symbolism with new depth. What is needed is a new pattern of rapport with the planet. Here we come to the critical transformation needed in the emotional, aesthetic, spiritual, and religious orders of life. Only a change that profound in human consciousness can remedy the deep cultural pathology manifest in such destructive behavior. Such a change is not possible, however, so long as we fail to appreciate the planet that provides us with a world abundant in the volume and variety of food for our nourishment, a world exquisite in supplying beauty of form, sweetness of taste, delicate fragrances for our enjoyment, and exciting challenges for us to overcome with skill and action. The poets and artists can help restore this sense of rapport with the natural world. It is this renewed energy of reciprocity with nature, in all of its complexity and remarkable beauty, that can help provide the psychic and spiritual energies necessary for the work ahead.

CHAPTER 4

Historical and Contemporary Spirituality

TRIBAL AND village peoples knew their local geographical region well, under the overarching sky, with the sun and clouds by day and moon and stars at night. This familiarity was expressed in mythic forms that accommodated personal experience and traditional environmental knowledge. This knowledge was different from contemporary science, with its emphasis on mathematical measurement and analytical precision. Now we have immense volumes of knowledge from science and technology that extend our vision and analytical skills and give us the precise measurement of everything, from the tiniest subatomic particle to the vast distances of the heavens. Whatever the advantages of this new human knowledge, it has made it difficult to maintain the needed physical and psychic rapport with the Earth we walk on, the local region we see with our eyes, the soil we cultivate for our food, the trees that surround us, the birds that sing in the evening, the flowers that bloom in the meadows where we collect hay for the animals we feed. Every organism must survive and grow in some immediate community, some intimate surroundings. Even every tree is acclimated to some particular region.

Humans also must settle into some particular cultural context, but we possess a special capacity of mind and imagination that enables us to settle into any area of the Earth and even, it seems, into any cultural tradition. In more recent times, thanks to our newly discovered means of travel and communication throughout the entire planet, we seem to need an even wider context for our lives. We have begun to realize some of the dreams of the ancient Stoics concerning the great city of the world.

A comprehensive human community is being created at the present time. Any vital spirituality of the present must be established within this perspective; otherwise, it will not express the realities of the human situation. If we are to fulfill the historical demands of our own existence, each person must, in accord with his or her abilities and opportunity, assume the global human heritage as a personal heritage.

If in former periods it was sufficient for us Westerners to live spiritually within the religious traditions of the West, this is no longer true. A fully adequate spiritual tradition can no longer be grounded simply within Western spiritual resources, even though individual traditions must be maintained and strengthened as the immediate context within which an individual grows and develops. For the continued existence and further development of the individual traditions, we also need a vital relation with other traditions.

This perspective is as necessary for spirituality as it is for other phases of human activity. All human activities now require development of their particular traditions within a global context, not an isolated one. Education is taking place within this perspective. Contemporary literature and the arts exist within a multicultural context. The languages of the world are being modified by the universal flow of thought. Political and economic life both function in this context. In every case, however, the more universal function also must be rooted within the local traditions. Any function that can be fulfilled more effectively within the local context must not be taken over by the more comprehensive. The village context is an important unit of human spiritual as well as economic development.

We must be mindful that the bioregions are not isolated fragments of the biosystems of the planet. Bioregions are relatively, not absolutely, self-sustaining. Rivers flow through a diversity of regions. Animals, especially birds, migrate through a sequence of bioregions. Air quality is determined by global processes, not by any single component of Earth's functioning. The atmosphere and the seas are both global commons.

The bioregions themselves are undergoing a continuing sequence of changes in relation to one another. Cultural forms are intimately related to the regions of their origins. Yet they are constantly being enriched by influences from the outside while at the same time sharing their own cultural treasures with others. Despite the conflicts that remain in the political, cultural, economic, and religious spheres, a differentiated yet comprehensive global human community is being established. Within this community, a comprehensive yet diversified spirituality is finding expression. This comprehensive spirituality must be considered as the affirmation, not the negation, of the individual traditions. The ideal is not a mixture of traditions but the further differentiation of the individual traditions within the comprehensive communion of all the traditions with one another.

While such change is taking place in a spatial-cultural extension of human activity, there is also a deepening of a temporal-historical awareness. This time/developmental process is the deeper force that is differentiating even as it is unifying the whole range of human consciousness. We now see ourselves as the most recent moment and the fullest expression of a consciousness that emerges out of distant geological ages. These earlier stages of human development have left an abiding impress in the depths of the human psyche, even as the various geological ages remain in the very structure of the Earth. So, too, many existing indigenous peoples maintain in the present earlier cultural forms even as they confront contemporary challenges and participate in modern technologies.

We humans, as the first true Earthlings with our own type of historical consciousness, now encompass the varied cultures of the Earth from East to West and from the Paleolithic to the present. Even before the Paleolithic, we experienced those awesome ages in which the life sequence emerged and spread over the globe in all its varied forms of expression. Beyond the four billion years of Earth's development, we experienced the formation of the galaxies and the background radiation from the formation of the universe itself. However vast and overwhelming these experiences of time and space, they are made manageable as soon as we bring them within the range of consciousness. At least in view of the unmeasured spaces of the universe, Earth reveals itself to us as a small planet. This experience is strengthened by the view of Earth as seen by voyagers to the moon. Even if our planet remains a realm of inexhaustible mystery, it has also become visibly situated in its basic dimensions as the comfortable home of the human family.

As part of this long cosmic process, the varied spiritual traditions scattered across the globe are not of the past, nor are they simply of the Earth. In some manner they were born when the galaxies appeared in the limitless swirl of space. The same dynamics at work in these ages have found unique expression in the formation of the Earth, with its myriad forms of life and their varied form of consciousness. In us, their human expression, the galaxies reflect on themselves in a special mode of awareness. The universe comes to know itself. Matter reaches its high transformation in those interior spiritual experiences wherein the universe comes to itself in its full identity and in its differentiated expression. In this ineffable mystery, every mode of being finds its peace and its communion with every other mode of being.

Of special importance is the influence on contemporary spirituality of indigenous peoples who understand with unique clarity the presence of the divine, the cosmic, and the human to one another. The distinctive spiritual experience of American Indian personalities seem destined to become a significant presence within the spiritual traditions of the human community. One example is Nicholas Black Elk (1863–1950) of the Oglala Lakota, whose life story is recorded by John Neihardt in the book *Black Elk Speaks* and in the collected field notes of Neihardt's research on Black Elk in *The Sixth Grandfather*, edited by Raymond DeMallie. There we read of Black Elk's experience of the moment when the song of the celestial stallion rang so vibrantly throughout the universe that "nothing anywhere could keep from dancing." He describes how "the virgins danced, and all the circled horses. The leaves on the trees, the grasses on the hills and in the valleys, the waters in the creeks and in the rivers and the lakes, the four-legged and two-legged and the wings of the air—all danced together to the music of the stallion's song."[1] We are beginning to experience our own participation in this comprehensive delight in existence so often described by the indigenous peoples of the world.

In addition to an awareness of the expansiveness of the globe and of the influence of the past that has emerged into consciousness, there is an increased awareness of the vastness of the future opening up before us. New possibilities for future phases of human transformation result from this understanding we have of ourselves and the universe in which we live. Compared to our knowledge of the past, our knowledge of the future is limited. But although we see only a little, we do perceive that the dimensions of the future must in some manner fulfill the expectations of both past and present. The rate of historical and cultural change has accelerated. Aspirations for the future pass beyond the dimensions of the past. Changes undreamed of before are now taking place in a brief span of decades—even within the limited range of future time available to our vision through conscious planning in the scientific and technological order.

Contemporary spiritual writers are generally preoccupied with one or another of these dimensions of human development. Some concern themselves with a defense and exposition of past traditions. Some are attracted to the spiritual developments that result from the convergence of traditions. Some are preoccupied with creating a spirituality suited to a united planet. Little has been done to bring these three concerns together. Yet none of these can be adequately dealt with in isolation from the others.

Before entering into a more detailed discussion of this issue, we might note the shock that passed over Western cultures in the decades following World War II. At the very moment when such expansive horizons of past, present, and future were opening up before us, we were cast into an inner anxiety and foreboding about ourselves and the meaning of it all. Writers such as Albert Camus (1913–1960),

Jean Paul Sartre (1905–1980), and Samuel Beckett (1906–1989) told us of the absurdity of all existence. The theater of the absurd developed in Eugene Ionesco (1909–1994) and Jean Genet (1910–1986).

Through the works of these writers, the sense of alienation once again found expression. Unable to bear the awesome meaning of life, we began to reject ourselves and the world around us. While indigenous peoples of the forests and deserts still have a sense of the magnitude of human existence and of our human capacity for living within divine and cosmic dimensions, contemporary Westerners are beset by a sense of confusion and alienation. Emotionally we show ourselves in the face of mystery less mature than indigenous peoples in the border deserts, island nations, and diminishing wildernesses of the world. The retrenchment of Western spiritual traditions and the consequent desolation we experience have been described with precision and fullness, for example in Beckett's *Endgame*.

Contemporary humans have no spiritual vision adequate for these new magnitudes of existence. In the very effort at discovery we abandon inner meaning. So far we have not been able to fill these magnitudes with a human presence such that we can really be comfortable with the world we live in. This art of comprehensive communion is a spiritual skill. To develop such a skill, to teach such a discipline, are the primary tasks of contemporary spirituality. We have lost the universe even as we walked on the moon. Recovering the universe requires an ability beyond that of maneuvering rocket engines or of engineering faster computers.

Stated briefly, we have lost the interpretative patterns of our existence, patterns generally designated as "myths." Myths are narratives that indicate the meaning of the human mode of being as well as the meaning of the universe itself. Our critical faculties, committed to the analytical processes of the rational mind, have destroyed the naiveté of ancient beliefs in favor of critical reflection and pragmatic realism. As the excitement of the new realism has diminished, we find ourselves encompassed by a world without meaning. There are only facts. The human itself is only another fact, qualitatively no different from any other fact of the world around us. The world of the sacred presented so forcefully in ancient myth and symbol no longer provides the atmosphere in which we can breathe humanly. Thus the suffocation of contemporary humans in consumerism and the excitement over the instant communication of the trivial.

Recovery of meaning involves a recovery of the sacred. That is the basic value that must first be identified and appreciated. This sense of the sacred requires recovery of ourselves, a return to the depths of our own being. We must in some manner manage the whole of existence in terms of the authenticity of our own deeper self. To accomplish this, we have to turn to three sources: our own past traditions, the spiritual insight of other traditions, and our present experience of the deeper realms of our own being.

When these sources are investigated, certain general orientations toward the world of meaning emerge. These are presented in the traditional myths and beliefs of the Western world, in the myths of the larger complex of world traditions, and in the symbols, stories, and dreams that emerge from the depths of primordial, intuitive modes of human consciousness. In these stories and symbols, similar things are expressed over such a wide geographic and historical area that a common human heritage begins to manifest itself. Many basic patterns and symbols of our spiritual development are similar across a vast diversity of traditions. Intercommunion of peoples in any significant degree must take place in and through these common symbols. Without these common symbols, very little can be achieved.

As regards our own society, the stage of demythologizing and deconstructing myth and symbol, which developed in the twentieth century, is over. Various disciplines offer new insight into the function of myth and symbol. Those who have penetrated deepest into the human psyche, whether in psychological analysis, in philosophical or religious studies, in the various spiritualities, or even in the understanding that science has of itself, now recognize the powerful and even determining role in human affairs played by ways of knowing beyond rational analysis. In this context, we can deal with the great paradoxes of reality and once again set up those needed spiritual disciplines upon which the human future depends.

Among these symbols, one that is most effective in understanding and responding to our present human needs is "the journey." This symbol offers a basis for unifying the past with the present, for bringing about a fruitful convergence of traditions, and for enabling the present generation to achieve a beneficial transition into the future. The idea of a journey is found in most societies that speak of their cultural patterns as a pathway to an authentic human mode of being. So we speak of the "way of life" as both the cultural and the spiritual form of a people.

This "way" encompasses the worldview of a society. It provides the interpretation of the universe, the experience of the human condition and the way of release from the human condition, the basic structure of values in the civilization, the legal and institutional structure, and the manner of human association by norms of kinship and community status. This way is depicted in the arts, it is the substance of the literature, and it forms the basis of education. Instruction in the higher mysteries of the way is incorporated into the initiation ceremonies whereby youths enter into mature participation in the life of the society. Those societies that developed a greater philosophical and theological awareness have established the way as an ontological absolute in terms of the *Tao* or the *Logos*. The spiritualities of the various societies were established according to this concept of a way as an interior journey to the real, a journey to the divine presence within.

The difficulty has been that so far in human history such a "way" has been ritualized in fixed and unchanging cosmic patterns. These seem not to allow for the type of historical change presently taking place. Yet these symbols will, if sufficiently examined, reveal themselves to be more adequate to the occasion than has generally been thought. These earlier symbols were never so totally fixed as would seem at first sight. They have been undergoing transformation within themselves from the time of their beginning. This is true within Hinduism, Buddhism, Confucianism, and the other traditions that have given shape and meaning to the human venture. All the traditions have been radically altered by cultural convergence in the past. None is a "pure" product of its own isolated society. Hinduism was modified by

Buddhism. Confucianism was modified by Buddhism. Islam in India was modified by Hinduism. Radical confrontation with the changes brought about by the new secular, scientific, and technological order is, indeed, a shattering experience for all traditional ways of life.

Even so, the ancient symbol of the journey is precisely the means needed to interpret our confusions and our confrontation with the devastating forces of the period. In reality, there is no devastation possible that is not confronted in principle by these narratives, in which we were brought up against the most awesome forces of destruction that the human imagination could conceive. The monsters of myth, such as Medusa, were not nursery tales for innocent children. They were confrontations with wild destructive forces of preternatural might.

Our modern journey into the future, along with the attendant agonies to be endured and perils to be undergone, is not the negation but the very substance of the ancient story. Earlier humans did not underestimate the problems that we must face in our individual or collective journeys through time. This can be seen in the journey of Gilgamesh, in the Homeric epics, and in the Exodus legend of the Hebrew peoples. It is found in the wanderings of the Buddha and in the travels of Confucius in China. It is expressed in the journey to India of the Chinese Buddhist monk Hsuan Tsang (c. 602–664), as narrated in the Chinese novel *Hsi Yu Chi* or *Journey to the West*. It is found in Augustine's story of the rise, progress through history, and termination of *The City of God* and in the journey of Dante in the *Commedia*. In all of these, external pilgrimage is the symbol of the journey of the individual soul, the particular society, and the human order itself into those interior depths wherein the sacred presence shines forth, all peril is surmounted, and final security is attained.

Without the symbol of the journey, it would be difficult to find meaning in our present venture through time, nor can we find the support we need for sustaining the sorrows and anxieties of life. So necessary is this narrative of the spiritual journey that only by establishing a new narrative can we engage these ancient tales of the meaning of life. The success of Marx for a century and a half was due to the journey symbolism outlined in the *Manifesto*. There all the ancient elements were brought together in a new presentation, one lacking the spiritual dimension of the earlier versions. Marx presents only the journey of matter through time and the inner dynamics of its evolution. There is no sense of the sacred, yet there is acceptance of the primary elements of the journey: the awakening to a strange and unsatisfactory setting of human existence, the need to seek a new form of life, the battles to overcome the destructive forces at work, and the final achievement of liberation, attaining the true self and the sacred paradise. All this is accomplished by human effort in alliance with transpersonal powers.

There is no need to stress the similarities with the Egyptian experience, with the Hebrew journey to the Promised Land, or with Buddha's experience of the destructive forces of age, illness, and death and deliverance. So too, the similarity to Dante's awakening in the Dark Wood and his journey through Hell and purgatory to a paradisal vision. In such Asian practices as Yogic and Buddhist meditation, the journey becomes almost totally interiorized. With Marx, the journey is almost totally exteriorized in terms of social tensions and their ultimate resolution in a peaceful world community. In both, however, the journey context is identifiable.

Of special significance is the cause or origin of the journey—the discovery that we exist in an unacceptable situation. The human must experience a transformation. This unacceptable situation involves not primarily the external surroundings in which we find ourselves but the unacceptable structure of our own being. This is what is involved in Dante's experience of the Dark Wood; in India's experience of *Duhkha*, of sorrow; and Gilgamesh's experience of mortality. While these experiences of the human condition are structured differently in the various traditions, certain essential characteristics are found in each. These are the unacceptable states in which humans discover themselves and need to transform. This spiritually transformed state of being is achieved not simply through personal effort but with the aid of more than human powers. We must enter into combat with demonic forces; we must endure affliction and even death prior to attaining an acceptable human situation. This situation is described in terms of paradise, of interior communion with the divine, of attaining sacred status, of immortality or emergence out of the merely temporal into the world of the eternal, of the final establishment of justice and peace within an integral human community, of the coming of the divine kingdom.

The journey is a common symbol found in various societies, and it is associated with another symbol also widespread, that of the Cosmic Person. The Cosmic Person is seen in the *Mahapurusha* of the Indian world, the Cosmic Buddha of the *Lotus Sutra*, the Sage in his identity with the entire order of things in the Chinese world, and the Cosmic Christ of Christianity. Awareness that humankind has a real, fundamental unity and that each of us in our full dimensions shares in this unity forms an effective basis for uniting peoples with one another and for relating the human world to the universe itself. This higher human personality appears in literature as "the hero of a thousand faces," in the words of Joseph Campbell (1904–1987). This higher personality is the Everyman of the medieval period, the Dante of the *Commedia*, and the Man-writ-large of Shakespeare (1564–1616).

So far, however, the symbol of Cosmic Person has been applied to the relation of humans to the universe and to human development through different periods rather than to the encompassing of the various cultural traditions within the metaphor of a single human. This latter has never been worked out satisfactorily, although it has found generalized expression in Pierre Teilhard de Chardin's (1881–1955) description of the rise of the noosphere and the convergence of peoples on a universal scale. In this manner, an all-inclusive higher personality is formed in preparation for a final unification at point Omega. While this symbol of the Cosmic Person has been a significant force in our past conceptions of ourselves, it may also provide a basis for understanding the various traditions as they now converge toward one another. This convergence of traditions, which seems destined to be an intensive experience in the future, needs spiritual interpretation of a high order. Because individuals will be unable to exist in isolation from social controls or social influence, the totality must be spiritualized within a meaningful way of life.

This comprehensive spirituality, which has reasserted the mythic as the means of communicating with the world of the sacred, finds itself in accord not only with the past course of human history and with the various cultures of the world but also with basic aspects of contemporary developments in the social order and with contemporary psychological understanding of the deeper self of the human community. This word "comprehensive," used in describing the spirituality needed by our contemporary world, also indicates one of the most significant themes of this discussion: the theme of totality. A spirituality suited to contemporary humans must rest on the drive we feel for a total experience of the universe. This is what propels us into the primordial past as into the distant future, into the outer dimensions of the universe as well as into the fantastic worlds hidden in the smallest particles of matter. We must walk on the moon both as a physical experience and as a mystical symbol of our inner journey. This drive toward fulfillment includes the quest to understand the deepest realms of the unconscious self as they are indicated by symbols revealed in dreams. This search into the deepest origins of psychic experience reveals that as humans we are centered in our place within the whole of things; the individual person seeks the reality of the whole and the whole of reality. This is the importance of the *mandala* symbol. Each of us must experience ourselves as both center and circumference. When we reach the integrative phase of our own personal experience, we also experience the integrative moment of the universe itself and of that supreme mystery in which the universe and the self exist. This is the achievement of the sacred wherein all oppositions are reconciled.

In some sense, the spirituality we need already exists and is being communicated to us by the larger human tradition. Indeed, the human tradition in all its multiform expression is the primary bearer and teacher of the spirituality we seek. This spirituality cannot be created anew. Human history cannot be set aside. The ancient symbols cannot be ignored. We must simply become conscious of the deeper and more universal forces at work in our own development. Spirituality is not something an individual or school of thought thinks up under some inner pressure to detach from the vulgar ways of humans so as to live in an esoteric realm of interiority. It is, rather, a profound expression of the mystery of participation in a total way of life, formerly of single cultures but now of the human community. The discovery of ourselves must include discovery of that spirituality which has supported and directed the human venture. This spirituality imposes itself just as inner creative powers impose a poetic vision that cannot be refused by the poet or the artist. This spirituality is not a future possibility but a present reality widely experienced but little understood.

This multiform spirituality can be observed at work within each society in the contemporary world. In and through its symbols, Western influences take root in Asia and Asian influences are apparent in the West; indigenous influences are strong among the more scientific-technological societies, and influences from these more commercial societies are found among indigenous peoples everywhere. Already we see that these traditions do not destroy one another or take away the distinctive characteristics that identify each one. Each keeps its original substance while bringing a new vitality to the larger context. Each lends support to the others in facing the historical crisis of scientism and secularism. Together the various traditions constitute a functioning public spirituality of the global village.

This is what is needed: a differentiated public spirituality for the human community. This spirituality in its historical depth and cultural diversity reaches its full expression when it enters into intimate communion with the vital forces of contemporary science and technology. The major proponents of these areas also now realize that their work acquires its meaning and value as well as discipline and limitation from the symbolic narrative within which it lives. The driving force for science and technology is also found in the symbol of the journey. When the scientists write the story of their discoveries of the universe, they have no way of describing it except in terms of the journey of matter through the various phases of transformation and the simultaneous journey of the human mind, wherein this journey of matter becomes conscious of itself and begins, in a new way, to guide itself into the future. This journey of matter and of mind is a single journey that can only be described as having two aspects, one expressed in physical terms, the other in terms of consciousness.

There is still another aspect of the journey symbol that should be mentioned, namely the death-rebirth symbol experienced in the form of periodic upheavals and transformations. Prior to the historical sequence of human development, the formation of the galaxies and the sequence of eruptions on the Earth often came in the form of vast upheavals and catastrophic movements, which shaped the continents and delimited the seas, raised the mountains and formed the valleys of the Earth. With its human expression, however, after the transition from the Paleolithic to the Neolithic period, the movements of Earth were fundamentally fixed; within this determined sequence of cosmic rhythms we established the rhythms of our own activities, accentuated by ritual celebration of the principal moments of cyclical and seasonal change, to bring the inner spiritual world into accord with the outer cosmic world. Only with the biblical experience could the death-rebirth cycle be seen more in radical historical transformation than in the seasonal sequence.

Death-rebirth has a special place in the liturgy of the Easter Vigil. Although this liturgy originally derived from the cosmic renewal rites that predate the Passover celebration, it has been adapted to the theme of historical renewal as this takes place through periodic transformations in the social-historical order. These are recalled in the Easter ritual in terms of the destruction of the world and its renewal at the time of Noah, the desolation of Egypt and the transition of the Hebrew peoples to the Promised Land, the Babylonian captivity and the restoration afterward. But more than the symbol of periodic historical renewal, there is the symbol instituted by the prophets of a total earthly renewal in the Day of the Lord, a symbol later developed into the period of the Reign of the Saints and the Descent of the Heavenly Jerusalem. These millennial symbols of the transformation and final healing of the human condition have given to Western societies our exceptional historical drive. This has been particularly true during periods of human suffering on an extreme scale. The vision of this Heavenly Jerusalem descending upon the Earth in total peace and abundance has given the human community a type of historical expectation such as we never had before.

This expectation has vastly increased the sensitivity we experience to our human condition and has made us more than ever desirous for total transformation both of ourselves and our environment. Evolutionary processes, which are often slow and laborious, have become intolerable. There must be immediate paradise. Thus there is the need not only to endure periodic historical convulsions but to bring these on by positive measures, much as the American Indians inflicted pain on themselves during their vision quests in the hope of bringing about the healing power of the Great Spirit through the intensity of their giving of themselves. In more recent centuries, this compulsive need for a shattering of social forms to introduce a reign of justice has expressed itself in the Western world on numerous occasions.

This problem of the gradual and the immediate, the evolutionary and the revolutionary, is the abiding tension in history. It is the cruel ambiguity in the prophetic enunciation of the Day of the Lord. The increasing acceleration of history has led to a growing sensitivity to the span of time to be endured before the day of bliss arrives; this in turn has led to the repeated triumphs of revolutionary movements over those more evolutionary methods, as can be seen in the triumph of the Bolshevik interpretation of Marx over the Menshevik interpretation. If total change is desired, all minor improvements are inadequate. There is no time to lose. Centuries of slow improvement will never lead to the alleviation sought and will not fulfill the biblical forecast of the healing of sorrow or the reign of the saints or the descent of the Heavenly Jerusalem. Nor does it seem that such an evolutionary process fulfills the requirements of the hero myth, which involves the engagement of demonic forces in a decisive combat that leads to the supreme treasure, the beautiful maiden, the authentic self, the divine presence.

As the future course of Marxism after 1903 was determined principally by the Bolsheviks, so the course of the human community seems to be determined by the more intense technological forces within the community. The causes of this go back to the millennial expectations of Western society that have been communicated to the human community on a wide scale. Movements such as the T'aiping Rebellion, the Cargo cults of New Guinea, the Ghost Dance religion of the American

Indian, the black power movements in American political life: all these are rooted in the prophetic announcement that justice will one day be achieved, that the human condition will be radically altered by a complete renewal of the established order of things in favor of a new Earth and a new heaven.

Since this new Earth has not come down to us from heaven, then we must bring this new age into being by violent efforts directed toward seizing control of the deepest genetic as well as the most powerful physical forces within the phenomenal world. Here, then, is the dominant source of Western aggressiveness that is manifested against the past, against all present establishments, against other peoples, and against the natural world itself in a technological conquest that will enable us to exploit the inner constitution of things in a tyrannical manner.

To cure this situation might be an impossible task were it not that the various spiritual traditions are highly resilient and enduring. Although neglected and abused in a thousand ways, they remain unalterably alive. Now these traditions suggest anew their remedy for the situation. As we seek to escape from the wasteland about us, we witness these ancient springs once again flowing with cool water capable of sustaining us on the next phase of our journey. Contemplative traditions are renewed, prayer is again a source of wisdom, and the healing power of silence is rediscovered. As the need for a more mystical relationship with the Earth becomes more widespread, education could become an initiation into a wisdom tradition rather than simply an acquisition of factual data. A comprehensive program of spiritual reconciliation begins to emerge, for now it is clear that to save ourselves we must save the totality of the human community as well as the totality of the Earth upon which we live. Above all, we begin to understand the deeper dimensions of the present historical crisis.

A more creative phase of the journey to our authentic self is resumed. Personalities of comprehensive vision and all-embracing human sensitivities emerge: those who assume the human heritage as their own heritage, who are capable both of intensive engagement in human affairs and of detachment proper to an authentic personality, who combine an understanding of the ages with profound insight, interest, and sympathy for the work of scientists and technologists. Within this context, the individual person engaging in his or her own interior journey feels the support and guidance handed down from the past in the multiple traditions of humankind. All this belongs to each person, for now indeed the spiritual heritage of humankind is the spiritual heritage of all. In a reciprocal manner, the individual can feel that in and through self-integration a healing comes to all. Thus the great journey is forwarded to the fullness of the real, and the interior self and the human community are brought into the divine presence.

In this manner, a person is in communion with humans everywhere. We no longer divide humans into "we" and "they"; we no longer think of an effective spirituality for one segment of the community. However sublime such a spirituality, it would not emerge from the integral human tradition or the integral divine revelation; it would be lacking in its proper center, its proper focus, and could doubtfully sustain itself in any ultimate sense. A broader communion is needed. The journey of the individual and the particular group can no longer be separated from the journey of the human community. Indeed, the individual is instructed and sustained by this universal pilgrim community, and the individual brings to this community the special presence, abilities, and insights that he or she is able to give. The more intimate and the more universal our communion, the more sublime the presence of the human, the cosmic, and the divine realms are to one another. In this manner we attain our authentic existence; an integral form of contemporary spirituality is established.

PART II

CHAPTER 5

The Spirituality of the Earth

THE SPIRITUALITY of the Earth refers to a quality of the Earth itself, not a human spirituality with special reference to the planet Earth. Earth is the maternal principle out of which we are born and from which we derive all that we are and all that we have. We come into being in and through the Earth. Simply put, we are Earthlings. The Earth is our origin, our nourishment, our educator, our healer, our fulfillment. At its core, even our spirituality is Earth derived. The human and the Earth are totally implicated, each in the other. If there is no spirituality in the Earth, then there is no spirituality in ourselves.

Not to recognize the spirit dimension of the Earth reveals a radical lack of spiritual perception. We see this lack of spiritual insight in the early European Americans' inability to perceive the spiritual qualities of the indigenous American peoples and their mysticism of the land. The opaqueness is in us, not in the Earth, for the Earth expresses an abiding numinous presence. Christianity's failure to recognize the Earth's spiritual qualities constitutes a significant challenge for us. Indeed, Christianity's failure led to aggression against the tribal peoples of this continent and the land itself with a destructiveness beyond calculation.

Clearly, the Earth as spiritual reality has been generally ignored by the religious-spiritual traditions of the modern West. This alienation from the spirituality of the Earth goes so deep that it is beyond our conscious mode of awareness. While there is recognition of the Earth in the Hebrew and Christian scriptures and liturgies, the dominant tendency is to see the Earth as a seductive reality that brought about alienation from the divine in the agricultural peoples of the Near East.

A redemptive spirituality that functions without regard for the larger human community or regard for the natural world that supports life is not likely to be effective in our secular world. Such a spirituality uses a rhetoric unavailable for our secular world, or, if it is available, it risks widening rather than closing the tragic inner division between the world of daily affairs and the world of divine communion. It cannot offer an adequate way of interpreting the inner life of the community in a rhetoric available to the community. Nor does it establish an understanding of that authentic experience in contemporary life that is oriented toward communion with creation processes. Indeed, it does not recognize that the context of any authentic spirituality lies in the creation story that governs the total life orientation.

In traditional Christian thought, creation is generally presented as part of the teaching on "God in himself and in relation to his creation." But this metaphysical, biblical, medieval, and theological context for understanding creation is not especially helpful in understanding the creation of the Earth and of the universe, as presented in scientific textbooks of Earth or life sciences as studied by students in elementary, high school, or college.

These textbooks initiate the child into an understanding of the formation and development of the Earth that has more continuity with later adult life in its functional aspect than does the catechetical story of creation taken from biblical sources. This scientific presentation of the formation of the universe in which the children live and find their place is crucial for their future spirituality. In our era, the school fulfills the role of the ancient initiation rituals that introduced children to society and to their human and sacred role in society. The tragedy is that the sacred or spiritual aspects of the initiation process are absent.

The student is told the marvelous story of the physical emergence of the universe, the Earth, and the human, but without reference to the larger *meaning* of this process.

It may be that the later alienation of so many young adults from the redemptive sacramental tradition is, in some degree, due to this inability to communicate to the child a spirituality grounded more deeply in creation dynamics. How different might it be if our religious traditions understood the spiritual significance of experiencing the galactic emergence of the universe, the shaping of Earth, the appearance of life and of human consciousness, and the historical sequence in human development.

In this sequence, the student might learn that Earth has, from its origin, an intrinsic spiritual quality. For too long this spiritual aspect of the creation story has been missing. This spirit dimension of the universe and of the planet Earth needs to be established if we are to have a functional spirituality. The issue is how to give the child an integral world. It is also the spiritual issue for the modern religious personality. Our most immediate task is to establish this new sense of the Earth and the functional role of humans within the Earth community. One might ask: how are we to do this?

To begin, we need to understand that the Earth acts in all that acts upon the Earth. The Earth is acting in us whenever we act. In and through the Earth, spiritual energy is present. This spiritual energy emerges in the total complex of Earth functions as each form of life is integrated with every other life form. Even beyond Earth, by force of gravitation, every particle of the physical world attracts and is attracted to every other particle. This attraction holds the differentiated universe together, enabling it to be a universe of individual realities distinct from, but intimately present to, one another. The universe is not a vast smudge of matter, some jellylike substance extended indefinitely in space. Nor is the universe a collection of unrelated particles. The universe is, rather, a vast multiplicity of individual realities with both qualitative and quantitative differences, all in spiritual-physical communion with one another. Individuals of similar form are bound together in their unity of form. The species are related to one another by derivation: the later, more complex life forms are derived from earlier, simpler life forms.

The first shaping of the universe was into great galactic systems of fiery energy that constitute the starry heavens. In these celestial furnaces the elements are shaped. Eventually, after some ten billion years, the solar system and Earth were born out of the particles cast out into space by exploded stars. So far as we know, Earth and its living forms constitute a unique planet. On Earth, both plant and animal life were born in the primordial seas some three billion years ago. Plants emerged upon the land some six hundred million years ago, land that the planet Earth shaped through geological upheavals that formed the continents, mountains, valleys, rivers, and streams. The atmosphere developed slowly. The animals came ashore a brief interval later. As these life forms established themselves over some hundreds of millions of years, the luxuriant foliage formed layer after layer of organic matter that was then buried in the crust of the Earth to become fossil formations with enormous amounts of stored energy. One hundred million years ago, flowers appeared, and the full beauty of Earth began to manifest itself. Some sixty million years ago, birds were in the air and mammals walked through the forest. Some of the mammals—whales, porpoises, and dolphins—went back into the sea.

Relatively recently, the diversity in the forms of life expression brought about the human mode of being. These first humans were wandering food gatherers and hunters until some ten thousand years ago, when they then settled into village life. This led through the Neolithic period to the classical civilizations, which flourished for the past five thousand years.

Approximately four hundred years ago, a new stage of scientific development took place, which brought about a human technological dominance of the Earth out of which we had emerged. This can be interpreted as Earth awakening to an intellectual understanding of itself. It has led to Earth taking a certain amount of control of itself in its human mode of being. The story of this scientific understanding of the universe, and the consequent ability to affect the functioning of the planet Earth, is among the most dramatic aspects of the Earth story.

It is difficult to understand or explain the aggressive attitude that then caused or permitted humans to attack the Earth with such savagery as we now witness. That it was done primarily by a Christian-derived society, and even with the belief that this was the truly human and Christian task, makes an adequate explanation especially challenging. Possibly it was the millennial drive toward a total transformation of the earthly condition that led us, resentful that the perfect world was not yet achieved by divine means, to set about the violent subjugating of the Earth by our own powers, in the hope that some higher or less vulnerable life would be attained.

While a more serene life is a positive goal, it must be acknowledged that the negative, even fearful, attitude toward the Earth resulting from the general hardships of life led to the radical disturbance of the entire Earth process. The increasing intensity of exploitation of the Earth was also the result of the rising expectations of Western peoples. Furthermore, the Darwinian principle of natural selection encouraged an attitude that each individual and each species struggles for its own survival at the expense of the others. Out of this strife, supposedly, the glorious achievements of Earth take place. Darwin had only a minimal awareness of the cooperative and mutual dependence of each form of life on the other forms of life. This is remarkable: he himself discovered the great web of life, yet he did not have a full appreciation of the principle of intercommunion.

Much more can be said about the conditions that permitted such a mutually destructive situation to arise between ourselves and Earth, but we must turn to suggest something of the new attitude that needs to be adopted toward Earth. What is needed is a new spiritual, even mystical, communion with Earth, a true aesthetic of Earth, a sensitivity to Earth's needs, a valid economy of Earth. We need a way of designating the Earth-human world in its continuity and identity rather than exclusively by its discontinuity and difference. We especially need to recognize the numinous qualities of Earth.

We might begin with some awareness of what it is to be human, the role of consciousness on Earth, and the place of the human species in the universe. While the traditional Western definition of the human as a rational animal situates us among the biological species, it inadequately expresses the role we play in the total Earth process. The Chinese, for example, define the human as the *hsin* of Heaven and Earth. The word *hsin* is written as a pictograph of the human heart. It can be translated by a single word or by a phrase that conveys both feeling and understanding. It could be translated by saying that the human is the "understanding heart of Heaven and Earth." Even more briefly, in this context, we can say that the human is "the heart of the universe." Yet another way to translate *hsin* is to say that we are "the consciousness of the universe" or "the psyche of the universe." Here we have a remarkable feeling for the fullest dimensions of the human, the total integration of reality in the human, and the total integration of the human within the reality of things.

We need a spirituality that emerges out of a reality deeper than ourselves, a spirituality that is as deep as the Earth process itself, a spirituality that is born out of the solar system and even out of the heavens beyond the solar system. For it is in the stars that the primordial elements take shape in both their physical and psychic aspects. Out of these elements the solar system and Earth took shape, and out of Earth, ourselves.

There is a triviality in any spiritual discipline that does not experience itself as supported by the spiritual as well as the physical dynamics of the entire cosmic-Earth process. Ultimately, spirituality is a mode of being in which not only the divine and the human commune with each other but through which we discover ourselves in the universe and the universe discovers itself in us. The Sioux Indian Crazy Horse (c. 1840–1877) invoked cosmic forces to support himself in battle. He knew that those forces resided in the depths of his being and he in them. He painted lightning upon his cheek, placed a rock behind his ear, an eagle feather in his hair, and the head of a hawk upon his head. The Sun Dance Ceremony of the Plains tribes of the Lakota and Crow also recognizes the power of these cosmic insignia. The entire universe participates. Dancers wear symbols of the sun, the moon, and the stars, both on their ceremonial dress and painted on their bodies. The world of living, moving things is indicated by the form of the buffalo cut from rawhide and hung on the centering tree, as well as by the eagle feathers used by the medicine leaders to heal. The plant world is represented by the cottonwood tree, a generative life presence set up in the center of the ceremonial circle. The fullness of spiritual energy in the multiplicity of life itself is represented by the circular form of the dance area. In this manner, the participants commune with those numinous cosmic forces out of which we were born.

This cosmic-Earth order needs to be supplemented by the entire historical order of human development. For example, when the Greeks went into battle, they drew energy from recalling historical events and depicting these on their shields. Virgil (70–19 BCE) devotes several pages to enumerating the past and future historical events wrought on the shield of Aeneas by Vulcan at the command of Venus, the heavenly mother of Aeneas. Homer similarly describes the shield of his hero Achilles.

Today we are in a new position where we can appreciate the historical and the cosmic as a single process. This is the vision of Earth-human development that will provide the sustaining dynamic of the contemporary world. We must nourish awareness of this vision. Our language and imagery need to acknowledge both the physical and psychic dimensions of this organizing force. It needs to be named and spoken of in its integral form. Just as we see the unified functioning of particular organisms, so too Earth itself is governed by a unified principle in and through which the total complex of earthly phenomena takes its shape. When we speak of Earth, we are speaking of a numinous maternal principle out of which all life emerges.

In antiquity, this mode of being or maternal principle of the Earth was often personified. For example, "Earth" designates a deity in Hesiod and in the Homeric hymns. It is expressed as Cybele in the eastern Mediterranean world and as Demeter in the Greek world. An exception is the Hebrew community, which distinguished itself by its monotheism, with a God seen as the creator of all things. Biblical revelation represents a basic antagonism between the transcendent deity, Yahweh, and the fertility religions of the surrounding societies. The basic effort here is to maintain an asymmetry in the relationship between the divine creator and the created. However, in the doctrine of the Madonna or the Divine Mother enunciated in later Christian history, there are many passages indicating that Mary was thought of as the Earth in which the True Vine is planted and which had been made fruitful by the Holy Spirit. Unfortunately, this identity of Mary with the Earth was never adequately developed in association with the doctrine of the Incarnation.

It is entirely possible that the dialectics of history required that the first direct human association with these unique historical individuals, the savior and his mother, had to develop before any adequate feeling for the mystique of the Earth could take place. Perhaps, too, a full development of redemption processes was needed before this new mode of human-Earth communion could find expression in our times. However, it is clear that a shift in attention is now taking place. The most notable single event bringing about this shift is the new and more comprehensive scientific account now available to us of our own birth out of the Earth. This story of the birth of the human mode of being was never known so well as now. After the discovery of the geological stages of Earth's transformation and the discovery of the sequence of life in ancient fossil remains by Louis le Clerc (1707–1788), James Hutton (1726–1797), and Charles Lyell (1797–1875) came Charles Darwin's (1809–1882) discovery of the emergence of all forms of life from primordial life forms. His *On the Origin of Species* (1859) describes the human appearance only out of the physical Earth. The French priest-paleontologist Pierre Teilhard de Chardin (1881–1955) saw the human emerging out of both the physical and the spiritual dimensions of the Earth.

It is a challenge for contemporary Earth studies to narrate the story of the birth of the human from our Mother the Earth. Once this story is told, it immediately becomes obvious how significant the title Mother Earth really is. Our long motherless period may be coming to a close; hopefully, the long period of our mistreatment of planet Earth is being terminated. If it is not terminated, if we fail to perceive not only our earthly origin but also our continuing dependence on our Earth-mother, then it will be due in no small measure to the redemptive and transcendentally oriented spiritualities that have governed our own thoughts, attitudes, and actions.

In this mother-child relationship, an essential, fundamental shift has taken place. Until recently, the child was taken care of by the mother. Now, however, the mother must be extensively cared for by the child. The child has grown to become an adult. The Earth-human relationship needs to undergo a renewal similar to that which occurs in the ordinary process of human maturation, where both child and mother experience a period of alienation, followed by a reconciliation period characterized by a new type of intimacy, a new depth of appreciation, and a new mode of interdependence. This period of reconciliation is the historical period in which we are now living. This new mode of Earth-human communion requires a profound spiritual context and a spirituality that is equal to this process.

In addition to a renewed awareness of Earth as mother, a second observation concerning our newly awakening sense of the Earth is that a new phase in the history of the Madonna figure of Western civilization has begun. The association of the Virgin Mother with the Earth may now be a condition to return Mary to a meaningful role. Her presence may also be a condition for overcoming our estrangement from the Earth. Because of our emphasis in the Western world on personhood, it is insufficient to see the Earth itself only as universal mother. It must be identified with a historical person in and through whom Earth functions. Phrases referring to Mary as the Earth are found throughout Western religious literature. These need to be retrieved and explored as a subject of utmost importance for our entire Earth-human venture. The Earth needs embodiment in a historical person, and such a historical person needs an identification with the Earth to fulfill adequately her role as divine mother. The medieval period of Western Christianity deserves special exploration in this regard. Few if any other civilizations were so deeply grounded in a feminine mystique as medieval Europe. A vital contact with this earlier phase of Western civilization is hardly possible without some deep appreciation of its feminine component. Thus we cannot fail to unite in some manner these two realities: Earth and Mary.

A third observation concerning the importance of the feminine or maternal principle is that the emergence of a new age of human culture will necessarily be expressed through the symbol of woman. This is because of the identification of woman with the Earth and its creativity. Woman and Earth, while differentiated, are inseparable. The fate of one is the fate of the other. This association is given in such a variety of cultural developments throughout the world in differing historical periods that it is hardly possible to disassociate them. Earth consciousness and woman consciousness go together. Both play an essential role in the spirituality of the human as well as in the structure of civilizations. An overly masculine mode of being has contributed to our alienation from the Earth, from ourselves, and from a truly creative man-woman relationship; it demands a reciprocal historical period in which not only a balance will be achieved but even, perhaps, a period of feminine emphasis.

A fourth observation is to note our new capacity for subjectivity, for subjective communion with the manifold presences that constitute the universe. In this we are recovering the more primal genius of humankind. For in our earlier years we experienced both the intimacy and the distance of our relation with the Earth and the entire natural world. Above all, we lived in awareness of a spirit world, a world that could be addressed in a reciprocal mood of affectionate concern. Nothing on Earth was a mere "thing." Every being had its own divine, numinous subjectivity, its self, its center, its unique identity. Every being was a presence to every other being. This gave rise to relational worldviews expressed in kinship as well as to Earth-based rituals, epic poetry, and the nature-inspired architecture of past ages.

Confucian thought gave the clearest expression to this intimacy of beings with one another in its splendid concept of *jen*, a word that requires translation according to context by a long list of terms in English: humaneness, love, goodness, human-heartedness, affection. All beings are held together in *jen*, as in the epistle by St. Paul (1 Colossians 1:17) where he notes that "all things are held together in Christ." Another perhaps even better analogy is in Newton's universal law of gravity, whereby each particle of matter attracts and is attracted to every other particle in the universe. The law of gravity indicates a mere physical force of attraction, whereas the universal law of attraction for the Confucians is a form of feeling identity.

For this reason, in Confucianism, there is the universal law of compassion. As the early Confucian thinker Mencius (372–289 BCE) suggested, this is especially observable in humankind, for every human has a heart "that cannot bear to witness the suffering of others." When the objection was made to the neo-Confucian Wang Yang-ming (1472–1529) that this law of compassion is evident only in human relations, Wang replied by noting that even the frightened cry of the bird, or the crushing of a plant, or the shattering of a tile, or the senseless breaking of a stone immediately and spontaneously caused pain in the human heart. This would not be, he tells us, unless there existed a bond of intimacy and even identity between ourselves and these other beings.

Recovery of this capacity for subjective communion with the Earth is a consequence and a cause of a newly emerging spirituality. Subjective communion with the Earth, identification with the cosmic-Earth-human process, provides the context in which we now make our spiritual journey. This journey is no longer only the journey of Dante (1265–1321) through the heavenly spheres. It is no longer simply the journey of the Christian community through history to the heavenly Jerusalem. It is the journey of primordial matter through its marvelous sequence of transformations, in the stars, in the Earth, in living beings, and in human consciousness. This journey is an ever more complete spiritual-physical intercommunion of the parts with one another, with the whole, and with that numinous presence that has manifested throughout this entire cosmic-Earth-human process.

CHAPTER 6

Religion in the Twenty-first Century

IF THE finest consequence of the first Parliament of Religions, held in 1893, was the recovery of a profound sense of the divine in the human soul through the leadership of Swami Vivekananda (1863–1902), the finest consequence of the second Parliament of Religions, held in 1993, should be the recovery of an exalted sense of the divine in the grandeur of the natural world. Vivekananda himself recognized that the locus for the meeting of the divine and the human must take place in the natural world if it is to survive in the human soul.

This concern for the visible world about us came to expression in only a few of the papers given at the 1993 parliament, even though the main issue confronting the future in all aspects of the human venture will be to achieve a viable relationship between humans and the natural life systems of the planet. Right now, the human is a devastating presence on the planet. While ostensibly humans are acting for their own benefit, in reality they are ruining the conditions for their own survival and well-being. This applies to both our physical and spiritual survival, since the inner world of the soul needs to be activated by experience of the outer world in all its grandeur.

The pathos of the present is that the human community has lost its capacity to interact creatively with the other components of the planet Earth. This includes the landscape of the planet; the nurturing qualities of the air, the water, and the soil; the energy flow that enables the dynamic powers of the Earth to continue their functioning; the life systems that are integrated in an immense complex of patterns beyond full human understanding.

While human beings have never had a comprehensive understanding of the mysteries manifest in the world about them, in former times they had, through their religious traditions, a capacity for being a creative presence within the ever-renewing sequence of life upon the Earth. In the closing years of the twentieth century and the dawn of the twenty-first, we seem to have lost this capacity. Instead, because of our population growth and technological power, we have become a deleterious presence throughout the planet. We thought that we were improving the human situation; in reality we were devastating human life along with all the other components of the Earth community. Just now we have begun to realize what we have done. We realize that a degraded outer world leads immediately to a degraded inner world.

A recovery of the sublime meaning of the universe could lead both to a greater intimacy of the human with the manifestation of the divine in the natural world and to a greater intimacy of the different religions among themselves. It becomes increasingly clear that humans have a common origin and a common destiny with every other component of the Earth community. We live on the same planet. We breathe the same air. We drink the same water. We share the same sunlight. We are nourished by the same soil. In all these ways we share a common spiritual mode of being and a common physical sustenance.

At this moment, this emphasis on the integrity of the natural world as a condition for the integrity of the inner spiritual world needs to be emphasized, because the relationship of humans with the natural world has deteriorated in a devastating manner in the hundred years between the 1893 and 1993 Parliaments of Religions. The disassociation of the human from the larger context of nature is the most significant change in human existence since the earlier parliament. It is also the most significant change in the biosystems and in the chemistry of the planet. We might even say that we are contributing to the termination of the Cenozoic Era (the last sixty-five million years) in the history of the Earth.

The only future really possible for humans or for the planet is a transition into an Ecozoic Era, when humans would be present to the planet in a mutually enhancing manner. This would introduce not simply a new era in human affairs but a new era of the planet itself. This new geobiological period is the condition for the integral functioning of the planet in all phases of its activities, whether these be biological, ecological, economic, cultural, or religious.

Religion, we must remember, is born out of the sense of wonder and awe of the majesty and fearsomeness of the universe itself. The great decline of religion in industrialized countries can be attributed in large part to the loss of an experience of the grandeur of the natural world, because of our newly acquired technological control over so many aspects of the natural world. At present, we are completely encompassed by the world of human artifice.

This alienation from the natural world deprives us of the immediacy and intimacy with the natural world that we observe in indigenous peoples the world over. In their immediacy with the natural wonders of the world about them, these peoples have an intimate relationship to the sacred as manifest throughout the planet. The world is attractive yet threatening, benign yet fearsome. Divine powers enable fruits, berries, nuts, and vegetation to come forth. These same powers bring the monsoon rains and the withering desert winds, the arctic chill, temperate warmth, and tropical heat.

These experiences evoke in the human soul a sense of mystery and admiration, veneration and worship. This is beyond what is sometimes thought of as nature worship. Recognition of the divine as manifested in nature can be found in the teachings of all the spiritual traditions of the world—even in the teaching of Saint Paul in the first chapter of his epistle to the Romans (Rom. 1:20).

Humans feel a need to integrate with these forces in order to survive and to fulfill their human role in the order of things. Ever present in the human soul is a certain anxiety due to the limited understanding of humans, the fragility of human powers, and a consciousness that if we fail in the fulfillment of our obligations in the order of the universe then the Earth will not provide the necessities that enable life to survive in any integral manner.

The very purpose of our religious rituals is to enable us to enter into the dramatic manifestation of the divine in a sacred world. This is why we have bodies capable of movement and sensation and minds with a capacity for ritual. Through our sacred ceremonies, we enter into the primordial liturgy of the universe itself.

The earliest human responses to the world around them grew more elaborate as the great ritual civilizations came into being some three thousand years ago in South Asia, East Asia, the Mediterranean, and the Americas. By this integration of human affairs with the regularities observed in the sequence of natural phenomena, humans obtained the spiritual power needed to confront the challenging nature of earthly life.

Restoration of this sense of the natural world as divine manifestation has a special urgency because of the devastation that we are presently causing to the natural world. The religious forces of the world, with their sense of the sacred, can evoke the psychic energies needed to transform a declining Cenozoic Era into the emerging Ecozoic Era. That the natural world is in decline is evident from the judgment of the foremost biologists of our times. When Peter Raven, director of the Missouri Botanical Gardens, addressed a professional society of scientists in a talk entitled "We're Killing Our World,"[1] this was not simply a biological statement, it was also a religious statement, for life and death are religious issues, especially when it is a question of life or death for the Earth itself in many of its major life systems.

For the first time in human history, we are responsible for the weakening of the ozone layer, the extinction of species on an unprecedented scale, and the degradation of the planet itself. With our growth as a species and our technological prowess, we are extinguishing life on a scale unknown to the planet since the decline of the Mesozoic Era some sixty-five million years ago. It was during the later years of the current

Cenozoic period that humans and all our human values, thought, and religious awareness came into being. The insights and values that have shaped human civilization for several millennia will be diminished on a devastated planet. To save the planet is to save ourselves.

In the future, the great work of every human profession and institution will be to bring the Ecozoic Era into being so that humans will be present to the Earth in a mutually enhancing manner. In this context, the religions will recognize that the natural world has from its beginning been a mystical as well as a physical reality. As the primary manifestation of the divine, the natural world is the primary sacred scripture and the primary sacred community. There could be no verbal scriptures unless first there were the cosmic scriptures. There can be no human community apart from the Earth community.

This sense of the sacred character of the universe has been lost to many peoples of the scientific-industrial world, because we were told that the universe we discovered was a process without inherent meaning or direction, a universe knowable only by quantitative measurement. Thus has the universe and our sense of the divine presence throughout the visible world diminished. This could happen because we had already abandoned the world. We were ready to believe what we were told.

Economists told us that a violent plundering of the Earth, as so much indifferent matter, would better our human existence. Philosophical realists told us that any appreciation of the mystical dimension of nature was a sentimental romanticism. Politicians told us that the way to controlling power in the world was to conquer our own territory and then exploit the territories of other peoples.

Whatever the past situation, an exciting future is opening up before us as the various religious traditions recover their integral relation with the dynamic forces of the Earth. Since the gathering of religious orders in Assisi in 1986, the religions of the world have begun a new period in their own activities. New hope arises on the North American continent as sustainable farming methods enable the land to recover its fertility and the grasslands are cared for by such persons as Wes Jackson. On the other continents, similar movements are emerging.

Wangari Matthai in Kenya has developed the Green Belt Movement, which is involved in replanting trees. There is the Chipko movement in India and the work of Vandana Shiva, who is providing both a critique of industrial agriculture and directions for sustainable agriculture. Aida Valasquez has a significant role in the ecological movement in the Philippines, as does John Seed in Australia. Mikhail Gorbachev is leading the Green Cross Foundation to assist the less wealthy nations to conserve their natural endowment. Everywhere persons with a profoundly religious dedication to the Earth are speaking out.

The plundering of the forests raises protests. Eroded soil is replenished through organic agriculture on community-supported farms. Rivers of the world are defended against further destruction by dams. Meadows and wildflowers once again come into bloom. Diminished shoals of marine life recover their abundance. As all these things begin to happen, there is hope that the springtime of the Ecozoic Era will soon appear.

There is evidence that we are recovering our experience of the divine as a pervasive presence throughout the natural world. We see an expression of this new intimacy with the Earth in the poetry of Mary Oliver and Gary Snyder. In music, Paul Winter has brought the songs of other living beings into his compositions, Carman Moore's *Mass for the Twenty-first Century* gives expression to this renewal of the ecosystems of the Earth, and Maia Aprahamian has composed a choral symphony in celebration of the Universe Story. An Ecozoic culture, inspired by the religious sensitivities proper to the new era, has begun throughout the world.

Similarly, a more comprehensive consciousness is appearing, one which recognizes that we form a single sacred society with every other member of the Earth community, with the mountains and rivers, valleys and grasslands, and with all the creatures that move over the land or fly through the heavens or swim through the sea. This new consciousness is evident when we, for example, insist on environmental impact statements before we venture into any construction project that affects the functioning of the life systems of the region. Even a river in Florida that was put into a concrete trench some years ago at a cost of thirty million dollars is now being set free again. The concrete basin into which it was channeled is being dynamited, to enable the river to spread out once more over its flood plain.

In the emerging Ecozoic Era, we experience the universe as a communion of subjects, not as a collection of objects. We hear the voices of all the living creatures. We recognize, understand, and respond to the voices of the crickets in the fields, the flowers in the meadows, the trees in the woodlands, and the birds all about us; all these voices resound within us in a universal chorus of delight in existence. In their work *Biophilia*, E. O. Wilson and Stephen Keller have emphasized the feeling of humans with the larger array of living beings.

We in industrial America are beginning to recognize that the human is a subsystem of the Earth system and that our first obligation in any phase of our human lives is to preserve the integral functioning of the larger world we depend on. We were brought into being in and through the Earth. We survive through our intimate presence to the Earth. While this is true in economics, governance, and the healing sciences, it is also true in religious affairs, since we are members of a single sacred community that includes every component of the earthly reality.

New religious sensitivities emerge as we understand better the story of the universe, which is now available to us through scientific inquiry into the structure of the universe and the sequence of transformations that have brought the universe, the planet Earth, and all its living creatures into being. This new scientific story of the universe has a mythic, narrative dimension that lifts this story out of a prosaic study of data to a holistic spiritual vision.

This new creation narrative enables us to enter into the deep mystery of creation with a new depth of understanding. It is our human version of the story that is told by every leaf on every tree, by the wind that blows across the fields in the evening, by the butterfly in its journey south to its winter habitat, by the mountains and rivers of all the continents of the Earth.

Through this story we understand with new insight how every component of the universe is integral with every other member of the universe community. To be is to contribute something so precious that nothing before or afterward will ever contribute that special glory to the created world. Through this story we learn something about how the primordial mystery of the universe brought the planet Earth into being as the most blessed of all the planets we know of. We learn how life emerged and took on such an immense variety in its forms of expression. We learn too how we were brought into being and guided safely through the turbulent centuries. In our contemplation of how tragic moments of disintegration over the course of the centuries were followed by immensely creative moments of renewal, we receive our great hope for the future. To initiate and guide this next creative moment of the story of the Earth is the Great Work of the religions of the world as we move on into the future.

CHAPTER 7

Religion in the Ecozoic Era

One of the most striking things about indigenous peoples is that traditionally they live in conscious awareness of the stars in the heavens, the topography of the region, the dawn and sunset, the phases of the moon, and the seasonal sequence. They live in a world of subjects, that is, a world of inner expression shared by every mode of being, not a world of objects. Nothing is without its identity, its dignity, its inner spontaneity. Everything has its sacred dimension, which must not be violated. Even today, despite material and social fragmentation stemming in part from the colonial encounter, many of the Australian aboriginal peoples still live in a universe that originated in dreamings and was peopled by spirit presences. They live in a topography shaped by song.

How different is this world from the world we live in. We hardly live in a universe at all. We live in a city or country, in an economic system, or in a cultural tradition. We are seldom aware of any sympathetic relation with the natural world about us. We live in a world of objects, not in a world of subjects. We isolate ourselves from contact with the natural world except in so far as we enjoy it or have command over it. The natural world is not associated with the very meaning of life itself.

The prevailing attitude toward the natural world is evidenced in the writing of a widely read columnist in a national magazine. Discussing ecological issues, he wrote:

> I am no enemy of the owl. If it could be preserved at no or little cost, I would agree: the variety of nature is a good, a high aesthetic good. But it is no more than that. . . . Nature is our ward. It is not our master. It is to be respected and even cultivated. But it is man's world. And when man has to choose between his well-being and that of nature, nature will have to accommodate. Man should accommodate only when his fate and that of nature are inextricably bound up . . . in either case the principle is the same; protect the environment—because it is man's environment.[1]

Such is the difference between our modern attitude of defiance toward the natural world and an indigenous attitude of reverence. I mention this because we are at a critical moment in the story of human relations with the natural world. The most basic issue of our time, I propose, is not divine-human relations nor even interhuman relations. The most basic issue of our time is human-Earth relations. We have disturbed the geological structure, the chemical composition, and the biological forms of the planet in a disastrous manner with our population explosion and technological power. We have closed down the creativity of the Cenozoic Era (the last sixty-five million years) and are ending a chapter of the geobiological history of Earth. Earth is now in a state of recession; its basic life systems have become disturbed, toxic, or are extinguished. The tragedy is that the Cenozoic has been a lyrical period in Earth's history. This was the period when Earth came to its full florescence. Trees, songbirds, flowering plants, marine life, tropical rainforests: all of these and more came into being during this era.

It was in such a setting, amid such awesome surroundings, that we as humans came into being and had our primordial experience of existence, an experience so overwhelming that our awakening into conscious awareness may well have been simultaneously an awakening to the divine. Quite early in our more developed mode of existence, surely within the past forty thousand years, we entered into the great liturgy of the universe through seasonal rituals. We celebrated the renewal of life in the springtime and gave thanks for the fruitfulness of life in autumn.

While the first expression of these liturgies most likely can be traced back to an even earlier period of human history, in the past ten thousand years they have been extensively elaborated. We see this in the architecture, pageantry, musical compositions, and great literary works created during the previous five-thousand-year period of human history. In these works, the urban literate civilizations gave expression to our experience of the universe as a manifestation of the divine.

Later, the Western world began its scientific inquiry into the origin, structure, and functioning of the universe. From these empirical observations, a new understanding of the universe has come into being. The Earth is no longer seen simply as moving in renewing seasonal cycles; it also has an emergent irreversible process that began some 4.5 billion years ago. Earth came into its present state through a long sequence of transformations that moved generally from simpler to greater complexity in structure, from lesser to greater manifestation of consciousness, and from lesser to greater freedom in action.

This knowledge enabled Western peoples to gain increasing control over the functioning of the planet, which led to a violent exploitation of the natural world. We intruded on the planet in a manner that was, at least at that moment, advantageous for humans but disastrous for the planet and its basic capacity for self-renewal. The conditions of life were eliminated for a vast number of species. The beautiful, natural landscape was reduced to scenes of utmost desolation. The remarkable dynamism that brought forth such resplendence and fecundity over the past sixty-five million years was terminated.

Is it not strange that there is such little awareness of what has happened, that there is such little concern manifested by the basic cultural institutions of our society, or such little guidance toward establishing a viable relationship between the human community and the natural world?

The damage to the planet was done, for the most part, by the generation that was born in the early decades of the twentieth century and has lived through the greater part of that century. My generation has done what no previous generation could do, because they lacked the technological power, and what no future generation will be able to do, because the planet will never again be so beautiful or abundant.

When I ask how this could have occurred, I answer that my generation has been insensitive to the natural world. Throughout the twentieth century, my generation has manifested little feeling with the natural world, and thus the natural world could not establish any spiritual or mystical mode of presence within our consciousness.

When the settlers came to North America, they saw the forest and the wilderness as a dark, even demonic, world. It was a world to be conquered and exploited. There was little sense or understanding of humans as integral members of a single sacred community composed of every mode of being upon the Earth. Only humans constituted the sacred community; only humans had rights.

This arrogance toward the forests and wildlife of this continent was noted by writers from the early days of settlement and especially by some late eighteenth- and nineteenth-century writers, such as James Fenimore Cooper (1789–1851), who, in his 1793 novel *The Pioneers*, recalling his childhood years in the Cooperstown region of New York, noted that the forests and wildlife of New York State were being devastated. William Strickland (1753–1834), in an account of his journey up the Hudson River in 1794 and 1795, wrote of the settlers in this area:

> The backwoodsman has got possession of the soil and fire and axe and are rapidly leveling the woods. The backwoodsman has an utter abhorrence for the works of the creation that exist on the place where he unfortunately settles himself. In the first place he drives away or destroys the more humanized Savage the rightful proprietor of the soil; in the next place he thoughtlessly, and rapaciously exterminates all living animals, that can afford profit, or maintenance to man, he then extirpates the woods that clothe and ornament the country, and that to any but

> himself would be of the greatest value, and finally he exhausts and wears out the soil, and with the devastation he has thus committed usually meets with his own ruin; for by this time he is reduced to his original poverty; and it is then left to him only to sally forth and seek on the frontiers, a new country which may again devour.... The day appears not too distant when America so lately an unbroken forest, will be worse supplied with timber than most of the old countries of Europe.[2]

Strickland's description is less dramatic than Herman Melville's (1819–1891) in *Moby Dick*, the story of the pathological assault of Captain Ahab on the Great White Whale.

Until recently, the prevailing view was that the North American continent must in some manner be reengineered and its power appropriated. Otherwise it was simply wasted. Not to dam the western rivers—the Colorado, the Columbia, the Snake, or the Tuolumne, which flowed through the Hetch Hetchy valley of California—was wasteful. Not to exploit the Tennessee with a series of dams was to refuse the power and water offered there. Not to soak the soil with chemical fertilizer was to deny ourselves an increased harvest. Not to pave the roads was neglect. Not to take the petroleum from the Earth was to reject a god-given opportunity for bettering human life, despite the fact that nature had stored the carbon in the petroleum and in the forests so that the chemical composition of the air and water and soil could be balanced in some effective manner. That humans had the right to do what they pleased was self-evident, especially as their population grew from two billion to six billion over the course of the twentieth century.

To explain such an attitude, it is not sufficient simply to go back to nineteenth-century industrialization, nor to Newtonian physics, nor even to Francis Bacon (1561–1626) or René Descartes (1596–1650). We must push our inquiry back into the anthropocentrism of the Hellenic world, back to the biblical world and the scriptural foundations of our Western life. We need to reflect on the juridic metaphor that underlies the entire concept of covenant when this concept is extended to include the Earth and all its living creatures. We also need to reflect on the more profound implications of the biblical emphasis given to our experience of the divine in a historical rather than cosmological manifestation. Beyond this, we should consider the effect of the primacy of an emphasis on redemption rather than an emphasis on creation in Christian thought in recent centuries, especially in post-Reformation theology.

Whatever might come of such reflection, we need to provide some understanding of why the devastation of the Earth has come about within a civilization formed predominantly by biblical-Christian religion and Greek humanism. These two forces, together with Roman legal and governmental structures and the ideals brought into Europe by the tribal peoples from the central Eurasian continent, might be considered as the constituents out of which came our Western civilization.

Whatever the sources of the immense devastation of the planet, none of these traditions has yet offered any adequate response. That our religious and humanist traditions, our educational programs, our jurisprudence, our economics, our commercial-scientific-industrial establishments and the other shaping forces of our society all contributed equally to our present situation might be too extreme a position to propose, but to note that none were able to prevent the destruction produced from within our Western civilization seems entirely valid. To say that all of these traditions have been excessively committed to an anthropocentrism also seems a proper conclusion. It could be said that they all favored processes that in some manner permitted, even if they did not actually lead to, our present disastrous situation.

As we enter the twenty-first century, we would do well to consider our way into the future. I propose that we need to go from the terminal Cenozoic to an emerging Ecozoic period, defined as that period when humans would be present to the Earth in a mutually enhancing manner. I prefer the term "Ecozoic" to that of "ecological," since this term enables us to place the coming geobiological period in its proper context: the sequence from the Paleozoic Era (from six hundred to 220 million years ago), to the Mesozoic Era (from 220 to sixty-five million years ago), to the Cenozoic Era (the past sixty-five million years), and now to the Ecozoic Era. This might now be accepted as the proper sequence in articulating the ages of the Earth and the context for historical interpretation in any aspect of our human cultural development. This view is supported in statements made by the biologists E. O. Wilson at Harvard and Peter Raven, director of the Missouri Botanical Garden.

To recover such a situation where humans would be present to the Earth in a mutually enhancing manner, I believe we must return to a sense of intimacy with the Earth akin to that experienced by many indigenous peoples of earlier times. This can be done through our new story of the universe, which is now available to us through empirical inquiry into the origin, structure, and sequence of transformations through which the Earth has come to its expression at the end of the twentieth century. We have finally realized that our modern knowledge reveals a universe with a psychic-spiritual and physical-material dimension from the beginning.

Articulating this story fully would be the supreme achievement of modern intelligence. Once we appreciate this transmaterial dimension of the universe, we will be able to understand that the human story is inseparable from the universe story. Then we can see that this story of the universe is in a special manner our sacred story, a story that reveals the divine particularly to ourselves, in our times; it is the singular story that illumines every aspect of our lives—our religious and spiritual lives as well as our economic and imaginative lives.

We will also be able to appreciate the primordial unity of origin of every being. Through this unity of origin, every being in the universe is kin to every other being in the universe. This is especially true of living beings of Earth, all of which have descended through the same life process. Through this sharing in a common story, we come to recognize our total intimacy with the entire natural world. An impenetrable psychic barrier is removed. We are no longer alienated objects but communing subjects.

We will now recognize that the universe itself is the only self-referent mode of being in the phenomenal world. Every other being, including the human, is universe referent. Only the universe is a text without a context. Every particular mode of being has the universe as context. In this manner, we circumvent the problem of anthropocentrism, which is at the center of the devastation we are experiencing. We recognize that in every aspect of our being, we are a subsystem of the universe system. More immediately, we are a subsystem of the Earth system.

We discover the Earth in the depths of our being through participation, not through isolation or exploitation. We are most ourselves when we are most intimate with the rivers and mountains and woodlands, with the sun and the moon and the stars in the heavens; when we are most intimate with the air we breathe, the Earth that supports us, the soil that grows our food, with the meadows in bloom. We belong here. Our home is here. The excitement and fulfillment of our lives is here. However we think of eternity, it can only be an aspect of the present. The urgency of this psychic identity with the larger universe about us can hardly be exaggerated. Just as we are fulfilled in our communion with the larger community to which we belong, so too the universe itself and every being in the universe is fulfilled in us.

We might say that the universe, throughout its vast extent in space and its long sequence of transformations in time, is a single multiform celebratory event. The human might be described as that being in whom the universe reflects on and celebrates itself and the deep mysteries of existence in a special mode of conscious self-awareness. Our human role is to enable the universe to reflect on itself in a special mode of consciousness.

In a corresponding manner, our individual human self is fulfilled in our family self, our community self, our Earth self, our and our universe self. In the Christian world, the believing person is fulfilled by our larger Christ-self, as Saint Paul tells us in his epistle to the Colossians: in Christ "all things hold together."[3] In Buddhism, the believing person is fulfilled through participation in the Buddha nature. We find a similar teaching in Confucianism. In fact, no one has given better expression to this unity of the universe than the Neo-Confucian thinker Wang Yang-ming (1472–1529), who taught the doctrine of the One Body of all things. As he expressed it: "Everything from ruler, minister, husband, wife, and friends to mountains, rivers, heavenly and earthly spirits, birds, animals, and plants, all should be truly loved in order to realize my humanity that forms a unity, and then my clear character will be completely manifested, and I will really form one body with Heaven, Earth, and the myriad things."[4]

It is the attraction of the particular self to the great self of our being that draws us so powerfully to inquire into, understand, and appreciate the stars in the heavens and the wonders of the Earth. Every mode of being is needed, for every being shares in the great community of existence. In this community of existence we discover our own fulfillment.

The comprehensive community is the supreme value in the phenomenal order. Each being receives its identity, honor, and value through its role in the universe. Within this larger universe, the planet Earth constitutes a single integral community. It lives or dies, is honored or degraded, as a single interrelated reality. As regards the future, it can be said quite simply that the human community and the natural world will go into the future as a single sacred community or we will both experience disaster on the way. Just as there is a single sacred liturgy that reaches from the utmost heights of the heavens to the farthest depths of the sea, as there is a single providence that reaches from end to end mightily and orders all things sweetly, so there is a single community of existence, a single universe community. The significance of the present is that we now know the universe in its physical origin and structure. We know the stages of its emergence into being through empirical knowledge gained by observational studies and a long effort at understanding the data we have gathered.

That the human is a subsystem of the Earth system is most clearly evident in the economic order. To advance the human economy by subverting Earth's economy is an obvious absurdity. Yet our entire commercial-industrial system is based on this absurdity. Until recently, little mention of this self-destructive aspect of contemporary economics was made in the scores of books written on economics and management.

So too in the area of governance. We begin to understand that there already exists a governance of the Earth, a governance too subtle for us to understand. This governance enables the Earth to bring forth the immense variety of its living forms that interact so intimately and extensively with one another. The well-being of each is fulfilled in the well-being of the whole. This governance has capacities far beyond anything humans are capable of. Yet this primordial governance remains the context into which we must insert our human governance. Our human governance needs to function within the context of Earth's governance, just as our economic functioning needs to be an extension of Earth's economy.

So we might also say of healing. Earth is a self-healing community just as it is a self-sustaining community and a self-governing community. There can be little hope for human healing except through the assistance of an integral natural world. When Earth becomes toxic, humans become toxic. We lose the only context in which we can hope for that vigorous mode of well-being that should be ours.

Our greatest difficulty stems from the inability of our religious traditions to accept our new way of telling the story of the universe, which derives from our empirically derived data. This sacred story has a special role to fulfill in this transition from the terminal Cenozoic to the emerging Ecozoic. This story enhances rather than negates the other sacred stories of the universe that over the years have guided the course of human affairs among indigenous peoples and the classical civilizations that have presided over the greater volume of human expression through the centuries. The effort to interpret our immediate experience of the universe simply through our scriptural data involves a serious distortion in our way of thinking. It subverts the very basis of our primordial experience of the divine in the manifestations offered us throughout the universal order of existence.

Even apart from this issue, we must ask whether these earlier traditions can, out of their own resources, provide adequate guidance for the task before us. Assuredly we cannot do without the guidance of these traditions of the past. They provide understanding and insight into the divine that is not available from the story of the universe we are presenting here. Yet we are faced with vast realms of knowledge and power that require the new range of understanding available to us from this new insight into the world that surrounds us.

Our Western religious studies, which I am primarily concerned with here, need to provide both the story and the dream. We need the story to understand where we are in the unfolding reality of the universe. We are, here on the planet Earth, faced with a disastrous situation akin to those two awesome moments at the end of the Paleozoic Era 220 million years ago, when some 90 percent of all living species were extinguished, and at the end of the Mesozoic Era some sixty-five million years ago, when another broad extinction occurred, which eliminated the dinosaurs. Both of these extinctions had different causes and consequences, since the life forms were less developed at the time. The current extinction is being caused by human action within a cultural tradition shaped in a biblical-Christian and classical-humanist matrix.

The tragic flaw in both traditions seems to be an anthropocentrism that has turned into a profound cultural pathology. We can accept that both have much to offer to the future shaping of the human venture and to a rethinking of the human venture and the Earth venture; that these can manage our present situation out of their own resources is more questionable.

The biblical story, however valid, however unique in what it offers, no longer seems sufficient to address the issues before us. We also need the story of our past and our dream of the future Ecozoic Era, for this coming era must first be dreamed. Through the dream comes the guidance, the energy, and the endurance we will need. The transition that is before us will cost an immense effort and require a wisdom beyond anything that we have known before.

Here it is necessary to note that the planet Earth will never again function in the manner that it has functioned in the past. Until the industrial period, the vast realms of Earth were magnificent—the luxuriance of the tropical rainforests, the movement of the great whales through the sea, the autumn color of the eastern woodlands. All this and so much else came into being apart from any human design or deed. We did not even exist when all this came to be. But for the foreseeable future, almost nothing will happen that we will not be involved in. We cannot make a blade of grass, but there is liable not to be a blade of grass unless we accept, protect, and foster it. Even uninhabited wilderness and the species that dwell there are affected by us and must now be protected by us. There is much healing that must take place throughout the planet, healing that will at times require our assistance—although for the most part, the natural world will bring about its own healing, if only we will permit it to function within the dynamics of its own genius.

In this context, we are moving from a theology of religion and an anthropology of religion to a cosmology of religion. This is the direction where, I think, religious studies will inevitably go in the future. In earlier times, our religious inquiry was theological: it was organized around questions concerned with the existence and nature of God and the relation of creatures to God. Later, our religious concerns were largely anthropological, ministerial, and spiritual, organized around such studies as the sociology and psychology of religion and the history of religions. In the immediate future, our religious concerns will, I believe, be more cosmological. They will be much more sensitive to the universe as the primary religious mode of being and to ourselves being religious through our participation in the religion of the universe. There will, I believe, be an emphasis on the planet Earth and on the universe itself as a single sacred community.

The natural world will once again become a scriptural text. The story is written not in any verbal text but in the very structure of the universe, in the galaxies of the heavens and in the forms of the Earth. These are phases in the great story that is the primary presentation whereby the ultimate mystery of things reveals itself to us. The sacred community will be recognized as including the entire universe.

The ritual expression of human presence to the divine will include not only the seasonal rituals whereby the human enters into the renewing cosmological order but also new rituals celebrating those moments of irreversible cosmological transformation that took place in the formation of the galaxies and in the supernova implosions that finally enabled the planet Earth and all those expressions of life and consciousness to come into being on the planet Earth. We will recognize cosmological and biological as well as historical and religious moments of grace.

Much else might be said here in dealing with this subject of religion and the academic study of religion in the Ecozoic Era, but they can at least be hinted at by saying that a new phase of religious studies may be developing, one aimed at overcoming our human and religious alienation from our larger, more comprehensive, sacred community of the natural world. This new discipline has a certain urgency, because if this alienation is not overcome, our other studies may, in the not too distant future, become irrelevant. Of one thing we may be sure: the human community and the natural world will go into the future as a single sacred community or we will both experience disaster.

This, then, is our challenge—to move from a purely human-oriented or personal-salvation focus in our religious concerns to one that embraces the universe in all its forms. This will require an immense shift in orientation, one that recognizes our emergence out of the long evolution of the universe and the Earth. The study of religion will begin to reflect this orientation in the cosmology of religion.

PART III

CHAPTER 8

The Gaia Hypothesis: Its Religious Implications

RECENTLY, A number of scientists have noted the remarkable capacity of Earth for unified homeostatic adjustment to a diversity of outer conditions. This argument for the "organic" quality of Earth has become known as the Gaia hypothesis, a name taken from an ancient Greek designation for the Earth Goddess. The Gaia hypothesis suggests that Earth is a self-regulating organism that has maintained the optimal temperature, atmosphere, and conditions for life.

As we develop these thoughts concerning the Earth, there is a need for a cosmology of Gaia as well as a biology of Gaia, since ultimately everything in the universe finds its context of interpretation within the universe. This cosmology of Gaia is especially necessary as a context for any religious interpretation of our subject. This is true because religious experience itself emerges out of the wonder that strikes the human mind as it experiences the inexplicable grandeur of the natural world.

We seldom think about the Earth itself in its distinctive aspects, because we are enclosed so intimately within its fields and woodlands or lost amid the commercial frenzy of our cities. We do speak about nature, the world, creation, the environment, and the universe, but generally in these broad inclusive terms, even when the planet Earth in its limited and distinctive aspects is foremost in our thought. Earth in its full spherical contours was never experienced by us directly until we were able to observe the planet from outside itself, in space.

Recently we have also come to know Earth within the context of a more comprehensive knowledge of the universe itself. We have begun to understand something of how the solar system was born out of the larger processes of the universe, how Earth and all its living forms took shape, and finally how we ourselves emerged into being. But even with such scientific knowledge, we often lack deep feeling for or understanding of the mystique of the Earth.

Even so, we still respond emotionally when the natural world impinges on our consciousness, in both its quiet and its more dramatic moments. The entire range of our poetry, music, and art resonate with the deep mysteries of existence experienced in the world about us. We are moved in the depths of our being by the serenity of the sea on a quiet evening or by the terrifying wintry storms that sweep across the North Atlantic. So too is the sharp aesthetic—even physical pain—we feel as we stand on some mountain height and look out over distant hills.

We have difficulty, however, with this sense of mystery, because we no longer understand the voices speaking to us from the surrounding world. Our scientific preoccupations and relentless commercial exploitation of the planet have left us with diminished sensitivity to the natural world in the deeper emotional, aesthetic, mythic, and mystical communication it is offering to us. We are so enclosed in our human world that we have almost completely lost our intimacy with the natural world. This has been described in a 2005 book by Richard Louv entitled *Last Child in the Woods: Saving Our Children from Nature-Deficit Disorder*.

If as children we become proficient in human language, we generally remain deaf to the multitude of languages of the natural world about us. We become socialized into the human community while becoming alienated from the larger society of living beings. Earlier in human history, humans considered themselves the totemic relatives of animals.

The powers of the universe were grandfathers and grandmothers. A pervasive religious rapport with the spirit powers of the natural world developed, and ritual enabled humans to enter into the grand liturgy of the universe. Seasonal renewal ceremonies brought humans into the rhythms of the solar cycle and the renewing splendor of the Earth. Special or sacred buildings were placed on coordinates identified with the position of the heavenly bodies, something we seldom think about nowadays.

This was a period of wonder and creativity. Everything possessed its own life principle, its own distinctive mode of self-expression, its own voice. Humans, animals, plants, and all natural phenomena were integrated within the larger community. As we are told by Henri Frankfort (1897–1954), in his treatise *The Intellectual Adventure of Ancient Man*:

> The fundamental difference between the attitudes of modern and ancient man as regards the surrounding world is this: for modern, scientific man the phenomenal world is primarily an "It"; for ancient—and also for primitive—man it is a "Thou. . . . " The ancients . . . saw man always as part of society, and society as embedded in nature and dependent upon cosmic forces. . . . Natural phenomena were regularly conceived in terms of human experience and human experience was conceived in terms of cosmic events.[1]

This continuity between the human and the cosmic was experienced with special sensitivity in the Chinese world. Human activities and court rituals were carefully coordinated with the cycle of the seasons. If summer music was played in the winter, the entire natural order was considered to be disrupted. The supreme achievement of the human personality in this context was to experience one's own being as "one body with Heaven and Earth and the myriad things." In the vast creative processes of the universe the human was "a third along with Heaven and Earth" as a primordial force shaping the entire order of things.

In these examples, a sense of the sacred dimension of the Earth is involved, a type of awareness less available from our traditional

Western religions. This lack of intimacy with the natural world was further extended when Descartes proposed that the living world was best described as a mechanism, because there was no vital principle integrating, guiding, and sustaining the activities of what we generally refer to as the living world.

Yet, strangely enough, a new sense of the sacred dimension of the universe and of the planet Earth is becoming available from our more recent scientific endeavors. The observational sciences, principally through the theories of relativity, quantum physics, Heisenberg's uncertainty principle, the sense of a self-organizing universe, and the more recent chaos theories have taken us beyond a mechanistic understanding of an objective world. We now know that there is a subjectivity in all our knowledge and that we ourselves, precisely as intelligent beings, activate one of the deepest dimensions of the universe. Once again we realize that knowledge is less a subject-object relationship than it is a communion of subjects.

We can now begin to appreciate the limitations of the analytical processes of our inquiry into the natural world. If formerly we knew by reductive processes that considered the particle as the reality and the whole as derivative, we now realize that we cannot know particles and their power until we see the wholes they bring into being.

If we know carbon simply as one of the 117 elements, then we have only minimal knowledge of what carbon is. To understand carbon, we must see its central role in molecules, megamolecules, in cellular life, organic life, sense life, and even in intellectual perception, because carbon in a transformed context lives and functions in the wide display of all the gorgeous plants and animals of the Earth as well as in the most profound intellectual, emotional, and spiritual experiences of the human. There is a latent spiritual capacity in carbon, just as there is a carbon component to our highest spiritual experience. This we experience within our own being as we become more conscious that the universe process, the Earth process, and the human process constitute a single unbroken sequence of transformations.

While the ancients had more highly developed sensitivities regarding the natural world in its numinous aspects and in its inner spontaneities, we are not without our own resources, which, properly appreciated, can lead us into a mode of intimacy with the natural world. If for a while we lost the poetry of the universe, this significantly changed when the astronauts came home stunned by the immensity and beauty of what they had experienced. Especially overwhelming was their view of the planet Earth from the regions of the moon, almost two hundred thousand miles distant. A new poetic splendor suddenly appeared in their writings, a poetry that emerged from the Earth. Astronaut Edgar Mitchell tells us:

> Instead of an intellectual search, there was suddenly a very deep gut feeling that something was different. It occurred when looking at Earth and seeing this blue-and-white planet floating there, and knowing it was orbiting the Sun, seeing that Sun, seeing it set in the background of the very deep black and velvety cosmos, seeing—rather, knowing for sure—that there was a purposefulness of flow, of energy, of time, of space in the cosmos—that it was beyond man's rational ability to understand, that suddenly there was a nonrational way of understanding, that had been beyond my previous experience.[2]

This experience, with all its romanticist overtones, which burst forth so spontaneously at the apex of our scientific-technological expertise, seems to arise out of a long repression, as if this cry of delight had been stifled over the past centuries. That this experience was so widely shared by the other astronauts speaks to the validity of the experience. Yet even such a dramatic episode would not be so impressive unless we knew that these men were deeply aware of the extensive scientific knowledge that we have gained of the universe and were speaking not out of simplistic emotion but out of an awareness of the fourteen billion years needed for the universe to bring into being the wonders they were seeing.

This sensitive experience of the universe and of the planet Earth leads us to appreciate the ten billion years required for the universe to bring Earth into existence and another four billion years for Earth to shape itself in such splendor. For our present Earth is not Earth as it always was and always will be. It is Earth at a highly developed phase in its continuing emergence. We need to see the sequence of earthly transformations as so many movements in a musical composition. In music, the earlier notes are gone when the later notes are played, but the musical phrase, indeed the entire symphony, needs to be heard simultaneously. We do not fully understand the opening notes until the later notes are heard. Each new theme alters the meaning of the earlier themes and the entire composition. The opening theme resonates throughout all the later parts of the piece.

Thus the origin moment of the universe presents us with a stupendous process that we begin to appreciate in its magnificence as it unfolds through the ages. The flaring forth of the primordial energy carried within itself all that would ever happen in the long series of transformations that would bring the universe into its present mode of being. The original moment of the universe in its primordial energies contained the undetermined possibilities of the present, just as the present is the activation of these possibilities. This primordial emergence was the beginning of the Earth story as well as the beginning of the personal story of each of us, since the story of the universe is the story of each individual being in the universe. Indeed, the reality inherent in the beginning could not be known until the shaping forces held in this process had brought forth the galaxies, Earth, its multitude of living species, and the reflection of the universe on itself in human intelligence.

After the universe's origin moment, a sequence of other transformational moments took place: the shaping of the first-generation stars within their various galaxies, then the supernova collapse of first-generation stars. These creative moments brought into being the entire array of elements. These in turn made possible the future developments throughout the universe, especially on the planet Earth, where the expansion of life needed the broad spectrum of elements for its full development.

The gravitational attractions functioning throughout the universe gathered the scattered stardust into this second-generation star we call our sun, and surrounding this star, its nine planets. Within this context, Earth began its distinctive self-expression, a groping toward its unknowable and unpredictable future, yet carrying within itself a tendency toward greater differentiation, a deepening subjectivity, and a more intimate self-bonding of its component parts.

Such wonder comes over us as we reflect on the Earth finding its proper distance from the sun so that it would be neither too hot nor too cold, shaping its radius so that it would be neither too large and (thus make the Earth more gaseous, like Jupiter) nor too small (and thus make the Earth more arid and rocky, like Mars). Then the Earth-moon distance was established so precisely—the moon was neither so close that the tides would overwhelm the continents nor so distant that the seas would be stagnant and life could not emerge.

Profound mysteries were taking place all this while, the most mysterious of which was this setting into place of the conditions required for the emergence of life and human consciousness. Principally through the work of James Lovelock and Lynn Margulis, we now understand in some detail that the story of life is so bound up with the story of Earth's geological structure that we can no longer think of Earth as first taking shape in its full physical form and then life somehow emerging within this context. The simultaneous shaping of its physical form and the shaping of its life took place in intimate association with each other. The living forms that appeared in the early history of Earth were among the most powerful forces shaping the atmosphere, the hydrosphere, and even the geological structures of the planet.

But while we need to understand the shaping power of living forms in the sequence of Earth's transformations, we must understand that living forms themselves were brought into being by the shaping power of earlier Earth development. Always there is this integral relationship between the earlier and the later. In the larger arc of this transformation process, the simpler forms are earlier, the more complex forms later, just as the simpler atomic elements took shape in the earliest moments of the universe and the more complex elements came later.

Much else might be said about this early phase of Earth's development, yet it is sufficient to note that each of these early occurrences in the life development of the planet were decisive. Each had to happen at precisely the opportune moment in the sequence of Earth's development for the planet to be what it presently is.

While perhaps incomplete, the narrative as given here presents in outline the story of the universe and of the planet Earth as this story is now available to us. This is our sacred story. It is our way of dealing with the ultimate mystery whence all things come into being. It is much more than an account of matter and its random emergence into the visible world about us, because the emergent process, as indicated by the geneticist Theodosius Dobzhansky (1900–1975), is neither random nor determined but creative, just as in the human order creativity is neither a rational, deductive process nor an irrational wandering of the undisciplined mind but the emergence of beauty as mysteriously as the blossoming of a field of daisies out of the dark Earth.

On Earth we find the fulfillment of the primordial tendency of the universe toward clearly articulated and highly differentiated entities. Earth astounds us with the vast differences between itself and the other planets. Each of the planets has its own distinctive mode of being, but these other planets are all much more like one another than any of them are to Earth.

This unique mode of Earth-being is expressed primarily in the number and diversity of living forms that exist on Earth, living forms so integral to one another and with the structure and functioning of the planet that we can appropriately speak of Earth as a "Living Planet." This term is used neither literally nor simply metaphorically but as analogy, somewhat similar in its structure to the analogy expressed when we say that we "see," an expression used primarily for physical sight but also used to connote intellectual understanding. A proportional relationship is expressed. The eye is to what it experiences as the intellect is to what it experiences. The common quality is that of subjective presence of one form to another as other. In this experience, the identity of each is enhanced, not diminished.

So in using this term "living" in speaking about a tree as a living being and in speaking about Earth as a living being, we are indicating that some of the basic aspects of life, such as the capacity for inner homeostasis amid the diversity of external conditions, are found proportionately realized both in the tree and in the comprehensive functioning of the planet. In the tree, as the primary analogue, we have the basic functioning of the life process through its beginning as a seed with its identifiable genetic coding, its absorption of the energies of the sun, and the flow of nourishment from its roots through its trunk to its leaves. Then there is the process of self-reproduction through its seeds. This process produces a certain continuing transformation of the surrounding atmosphere, whereby the presence of the life process can be discerned.

So too Earth comes into being. Not, however, with an identifiable genetic coding guiding Earth through its stages of development to its maturity nor through birth from a prior Earth or living organism with the capacity to continue this generative process. Earth cannot reproduce itself. Yet notwithstanding, there are similarities that justify the use of the term "living" to describe Earth in its integral functioning, especially in its capacity for inner self-adjustment to the diversity of external conditions to which it is subject. This "feedback" process is so remarkable that, along with the capacity of the planet to bring forth such an abundance of life forms, Earth can be described not simply as living but as living in a supereminent manner.

The use of metaphor and analogy does not diminish the reality of what is being said. The more primordial realities can only be spoken of in a symbolic manner. To indicate that Earth is not exactly a living reality in the sense that a bird or a flower is a living reality is not to diminish the significance of Earth as a living being. It is rather to heighten the significance of what we are saying. Earth makes possible all those multiple forms of life upon the planet, not simply some single life form. Earth "flowers" into the immense variety of species, not simply into another Earth.

The deepest mystery of all this is surely the manner in which these forms of life, from the plankton in the sea and the bacteria in the soil to the giant sequoia or to the most massive mammals, are ultimately related to one another in the comprehensive bonding of all the life systems. Genetically speaking, every living being is coded not only in regard to its own interior processes but in relation to the entire complex of earthly being. This is to be alive and to be the fertile source of life.

Earth's fertility in bringing forth life is found in our reference to Earth as Universal Mother, as Gaia. Earth is Mother who gives birth to all the living forms that exist on Earth. However, these living forms have influenced the shaping of Earth, and they themselves derive from the period prior to the appearance of organic life on the planet. Thus it is Earth itself that is the subject most deserving of a maternal designation, not the biosphere. Earth is the larger subject that activates its being in the total complex of spheres that constitute Earth: the geosphere, the hydrosphere, the atmosphere, and the biosphere. None of these has existence or function apart from its unity in Earth.

We need to think of the planet as a single, unique, articulated subject to be understood in a story both scientific and mythic. Just as a tree is a unified subject capable of coordinating the vast diversity of activities involved in its emergence from a germinating seed, sending roots down into the soil, raising up its trunk, branching out in all directions, sprouting leaves, and finally scattering seeds for the further expansion of its life, so also Earth is a subject capable of coordinating the variety of activities whereby the various species come into being. In both instances, we are dealing with realities that need scientific and mythic modes of understanding.

The great benefit of the Gaia hypothesis is that it is an effort at a larger pattern of interpretation. It might be suggested, however, that the Gaia designation does not go far enough in this larger sense of the planet Earth as a "living" reality. Neither biological nor chemical studies alone can deal adequately with the superb achievement of Earth in its self-shaping from the beginning. Nor, apparently, do these give adequate consideration to the prior conditions for the appearance of life, which Earth, in its primordial phases, brought into being. As with so many basic concepts in the array of human knowledge, the concept of Gaia is multivalent, giving rise to an extensive range of development in a variety of disciplines. Here we are concerned with the insight into the dynamism of Earth as this appears within a more comprehensive cosmology.

The planet Earth might well be the most unique reality in the universe precisely in its capacity for bringing forth in the unity of a single being all those various modes of physical structure, organic life, and consciousness that presently constitute the reality of the planet. It also seems that Earth has the status of a privileged planet not simply within our solar system but possibly throughout the entire universe. This privileged status is especially evident in the conviction of some scientists that the universe is as old as it is and as big as it is because it takes a universe this old and this big to produce a planet such as Earth, which has the requisite conditions for the emergence of life and the human form of consciousness.

In this more comprehensive understanding of Earth we might recall the primordial tendency of the universe toward the communion of every being with every other being in the universe. Ultimately this brings us back to the curvature of space, the primordial expression of the comprehensive bonding force of the universe. This bonding, expressed in gravitational attraction, is a primary psychic-spiritual as well as a physical ordering principle of the universe's larger dimensions. Gravitational attraction keeps the divergent forces of the original emergence within the limits needed for the creative processes that have taken place over the centuries.

These two opposed forces, divergent and convergent, associated with the original emergence give us the curvature of space, a curvature sufficiently closed to hold all things together in an ordered universe yet sufficiently open to permit the creative process to continue over the centuries. It is this curvature that brings every component of Earth into intimate association with every other component. Everything within this curvature has not only its individual mode of being but its universe mode of being, since the universe is integral with itself throughout its entire extension in space and throughout the full sequence of its transformations in time. Indeed, nothing can be itself without everything else. Everything exists in multiple dimensions. A tree is a physical being, a living being, an Earth being, and a universe being.

For the human especially, these multiple modes of our being require both the activation of the physical and biological modes of our being and the activation of the psychic mode of our being. We have our individual self, our biological self, our Earth self, and our universe self. It is through attraction to the larger modes of our self that we are drawn so powerfully toward our experience of the Earth. We seek to travel throughout the Earth, to see everything, to experience the grandeur of the mountains, to plunge into the sea, to raft the rivers, to fly through the air, even to go beyond Earth into space. We seek this for the expansion of our being, even more than for the physical thrill. In all these experiences we come to know the further realms of ourselves and experience the deepest mysteries of existence—what might well be considered the numinous origins whence the Earth and the entire universe derive, subsist, and have their highest mode of fulfillment.

Thus the scientist seeks to understand Earth in all its geological and biological forms, to examine the inner realms of the atomic and subatomic worlds. Even recent concerns for understanding Earth as a living organism arise not from an arbitrary feeling that it would be an interesting venture of the human mind. We are, rather, impelled to this inquiry through our efforts at our own self-discovery. It is a mystical venture, for its ultimate purpose is to achieve a final communion with that ultimate reality whence all things come into being. The dedication of personal effort, the life discipline, the excitement of the discoveries made, the differences, the identities, the coherences, the moments of intellectual impasse—all these reveal a new form of religious enchantment and a quest for further revelatory experience. For the universe whence we emerged is constantly calling us back to itself. So too Earth is calling us back to itself, and not only to us but to all its components, calling them into an intimacy with one another and to the larger community within which all earthly realities have their existence.

Thus the larger explanation of any part of the universe is the cosmological explanation. So too it is with Earth. We need a Gaia hypothesis. But we also need a cosmological context for understanding the meaning of Earth as Gaia. This cosmological context is especially important for any consideration of the religious implications of the Gaia hypothesis.

Indeed, our scientific inquiry in this direction establishes the basis for a new type of religious experience different from but profoundly related to the religious-spiritual experience of the earlier shamanic period in human history. Since religious experience emerges from a sense of the awesome aspects of the natural world, our religious consciousness is consistently related to a cosmology that tells us the story of how things came to be in the beginning, how they came to be as they are, and the role of the human in enabling the universe in its earthly manifestation to continue the mysterious course of its creative self-expression.

From a religious perspective, we might consider that because of the diversity of life expression that is held together in such intimate unity, the Earth is a special presentation of the deep mysteries of existence whence religious consciousness arises. Thomas of Aquinas refers to "difference" as "the perfection of the universe." The reason is that the divine could not imagine itself in any single being, so the divine brought into being an immense variety of beings. Thus the perfection lacking to one would be supplied by the others. "Consequently the whole universe together participates in the divine goodness more perfectly, and manifests it better than any single being whatever."[3]

We could adapt this passage by simply saying that the deep mysteries of existence are manifested more perfectly in accord with the greater diversity held in the greater unity. This provides us with a way of dealing with the special role of the Earth as revealing the deepest realms of existence with a perfection unequalled in any other mode of being we know of. For in the Earth we have our most magnificent display of diversity caught up into the coherence of an unparalleled unity.

In this context, we can understand the special numinous quality attributed to the Earth. In its own self-manifestation, the Earth is also a revelation of the ultimate mystery of things. The sense of awe and mystery that was evoked in the earliest human awakening to the universe is beginning to awaken once more within this new context of scientific understanding. We have indeed lost contact with the world of the sacred, as this sacredness was experienced through a spatial mode of consciousness in which time was perceived to move in eternally recurring seasonal cycles. Yet we now begin to experience the sacred dimension of our new story of the universe as an irreversible emerging process.

No longer are we celebrating simply the seasonal renewal of the living world. We now are experiencing in the world around us the primordial emergence of the universe in the full surge of its creativity. We are integral with the process. We experience the universe with the delight of our postcritical naiveté.

Never before have any people carried out such an intensive meditation on the universe and on the planet Earth as has been carried out in these past few centuries in our Western scientific venture. Indeed, there is a mystical quality in the scientific venture itself. This dedication, this sacred quest for understanding and participation in the mystery of things, is what has brought us into a new revelatory experience. While there is no need for us to be professional scientists, there is an absolute need for us to know the basic story of the universe and of the planet Earth as these are now available to us by science.

CHAPTER 9

The Cosmology of Religions

THE UNIVERSE itself is the primary sacred community. All religious expression by humans should be considered participation in the religious aspect of the universe itself. We are moving from the theology and the anthropology of religions to the cosmology of religions. Throughout the twentieth century in America, there was an intense interest in the anthropology of religions, particularly the sociology of religions, the psychology of religions, the history of religions, and comparative religion. Because none of these forms of religious consciousness has been able to deal effectively with the evolutionary story of the universe or with the ecological crisis that is now disturbing Earth's basic life systems, we are being led to a cosmological dimension of religion both by our efforts at academic understanding and for practical issues of physical survival on a planet severely diminished in its life-giving capacities.

What is new about this mode of consciousness is that the universe itself is now experienced as an irreversible time-developmental process, not simply as an eternal, seasonally renewing universe. We are focused not so much on cosmos as on cosmogenesis. Now our knowledge of the universe comes primarily through our empirical, observational sciences rather than through intuitive or deductive reasoning processes. We are listening to Earth tell its story through the signals that it sends to us from outer space, through the light that comes to us from the stars, through its geological formations, and through the vast amount of data that the biosystems of Earth give us.

In its every aspect, the human is a participatory reality. We are members of the great universe community. We are not on the outside looking in; we are within the universe, awakening to the universe. We participate in its life. We are nourished by this community, instructed by this community, governed by this community, and healed by this community. In and through this community we enter into communion with that numinous mystery whence all things depend for their existence and their activity. If this is true for the entire universe, it is especially true given our human dependence on Earth.

From observable scientific evidence, we understand the story of the universe as an emergent process with a fourfold sequence: the galactic story, the Earth story, the life story, and the human story. Together these constitute for us the primordial sacred story of the universe.

The original flaring forth of the universe carried the present within its fantastic energies just as the present expresses those primordial energies in their articulated form. This includes all of the aesthetic, psychic, and spiritual developments that have occurred across the centuries. The universe, in its sequence of transformations, carries within itself the comprehensive meaning of the phenomenal world. In recent secular times, this meaning was perceived only in its physical expression. Now we perceive that the universe has been a spiritual and a physical reality from the beginning. This sacred dimension is especially evident in those mysterious moments of transformation the universe has passed through during its fourteen billion years of existence. These are moments of great spiritual and physical significance: the privileged moments in the Great Story. The numinous mystery of the universe now reveals itself in a developmental mode of expression, a mode never before available to human consciousness through observational processes.

Yet this seems not to mean much to our contemporary theologians. They remain concerned with scriptural interpretation, spiritual disciplines, ministerial skills, liturgy, the history of Christianity, the psychology of religion, and religious pedagogy. None of these areas of study has a direct concern for the natural world as the primary source of religious consciousness. This is one of the basic reasons why both the physical and spiritual survival of the Earth has become imperiled. Presently, we in the West think of ourselves as passing into another historical period, in a continuation of the long series of historical transformations that have taken place in the past and that are continuing on into the future. This perception is understandable. If we think, however, that the changes taking place in our times are simply another moment in the series of transformations that passes from classical-Mediterranean times through the medieval era to the industrial and modern periods, then we are missing the real magnitude of the changes taking place. We are, in fact, at the end of a religious-civilizational period. By virtue of our new knowledge, we are changing our most basic relations to the world about us. These changes are of a unique order of magnitude.

Our new acquaintance with the universe as an irreversible developmental process can be considered the most significant religious, spiritual, and scientific event since the emergence of the more complex civilizations some five thousand years ago. At the same time, we are bringing about a devastation of Earth such as the planet has never experienced in the four and a half billion years of its formation.

We are changing the chemistry of the planet, we are disturbing the biosystems, and we are altering the geological structure and functioning of the planet, all of which took some hundreds of millions and even billions of years to bring into being. This process of closing down the life systems of the planet is making Earth a wasteland, and we hardly realize that with each species of life on Earth we lose we also lose modes of divine presence, the very basis of our religious experience.

Because we are unable to enter into the new mystique of the emergent universe, we are unable to prevent the disintegration of the life systems of the planet taking place through the misuse of that same scientific vision. Western religions and theologies have not yet addressed these issues nor established their identity in this context. Nor have other religious traditions been any more successful. Mainstream religions have simply restated their belief and their spiritual disciplines in a kind of fundamentalist pattern.

We cannot resolve the difficulties we face in this new situation by setting aside the scientific venture that has been in process over these past two centuries, especially during this twenty-first century. It will not go away. Nor can we assume an attitude of indifference toward this new context of earthly existence. It is too powerful in its total effects. We must find a way of interpreting the evolutionary process itself. If interpreted properly, the scientific venture could even be one of the most significant spiritual disciplines of these times. This task is particularly urgent, since our new mode of understanding is so powerful in its consequences for the very structure of the planet Earth. We must respond to its deepest spiritual content or else submit to the devastation that is before us.

I do not consider that fundamentalist assertions of our former traditions can themselves bring these forces under control. We are not engaged simply in academic inquiry. We are involved in the future of the planet in its geological and biological survival and functioning as well as in the future of our human and spiritual well-being. We will bring about a physical and spiritual well-being of the entire planet or there will be neither physical nor spiritual well-being for any of our earthly forms of being.

The traditional religions have not dealt effectively with these issues nor with our modern cosmological experience because they did not originate in nor were they designed for such a universe. Traditional religions have been shaped within a dominant spatial mode of consciousness, that is, a mode of consciousness that experiences time as a renewing sequence of seasonal transformations. Although Judaism, Christianity, and Islam have a historical-developmental perspective in dealing with the human process, they lack an awareness of the development of the universe itself. They seem to have as much difficulty as any other tradition in incorporating an understanding of the developmental character of the universe.

Although antagonism toward an evolutionary universe has significantly diminished in Christian theology, our limitations as theologians in speaking the language of this new cosmology is everywhere evident. If much has been done in process theology in terms of our conceptions of the divine and the relations of the divine to the phenomenal world, this has been done generally in the realm of systematic theology. Little has been done in the empirical study of the cosmos itself as religious expression.

To envisage the universe in its religious dimension requires that we speak of the religious aspect of the original flaming forth of the universe, the religious role of the elements, and the religious functioning of the Earth and all its components. Since the human in its religious capacities emerges out of this cosmological process, then the universe itself can be considered the primary bearer of the religious experience. As Thomas Aquinas (1225–1274) tells us: "The order of the universe is the ultimate and noblest perfection in things."[1] Although Thomas was thinking of the universe in terms of its renewing seasonal sequence and not as an emergent sequence of irreversible transformations, the same principle applies. The universe is the primary referent in all human understanding.

This way of thinking about the emergent universe provides a context for the future development of the world's religious traditions. Indeed, the peoples of the world, insofar as they are being educated in a modern context, are now able to identify themselves in time and space in terms of the universe as this is presently described by our modern sciences. The problem is that they are not learning the more profound spiritual and religious meaning also indicated by this new sense of the universe.

This story of the universe is at once scientific, mythic, and mystical. Most elaborate in its scientific statement, it is among the simplest of creation stories. Most of all it is the story that we learn from the universe itself. We are finally overcoming our isolation from the universe and beginning to listen to the universe's story. If until recently we were insensitive in relation to its more spiritual communication, this is no longer entirely true. In this understanding, we have an additional context for understanding all the religious traditions, just as our more recent cosmologies do not negate but add to the Newtonian worldview and enable us to deal with questions that could not be dealt with in the Newtonian context. Now we have additional depth of spiritual understanding through our listening to the universe in ways that were not available through our traditional insights. Just as we can no longer live simply within the physical universe of Newton, we can no longer live spiritually in any adequate manner simply within the limits of our earlier religious traditions.

The first contribution this new perspective on the universe makes to religious consciousness is the sense of participating in the creation process itself. We bear within us the impress of every transformation through which the universe and the planet Earth have passed. The elements out of which Earth and all its living beings are composed were shaped by supernovae. We passed through the period of stardust dispersion resulting from this implosion-explosion of a first-generation star. We were integral with the attractive forces that brought those particles together in the original shaping of the Earth. We felt the gathering of the components of the earthly community and experienced the self-organizing spontaneities within the megamolecules, out of which came the earliest manifestations of the life process and the transition to cellular and organic living forms. These same forces that brought forth the genetic codings of all the various species were guiding the movement of life toward its expression in human consciousness.

This journey, the sacred journey of the universe, is the personal journey of each individual. We cannot but marvel at this amazing sequence of transformations. No other creation story is more fantastic in its account of how things came to be in the beginning, how they came to be as they are, and how each of us received the special characteristics that give us our personal identity. Our reflexive consciousness, which enables us to appreciate and to celebrate this story, is the supreme achievement of our present period of history. The universe is the larger self of each person, since the entire sequence of events that has transpired since the beginning of the universe was required to establish each of us in the precise structure of our own being and in the larger context in which we function.

Earlier periods and traditions also experienced intimacy with the universe, especially in those moments of cosmic renewal that took place periodically, mostly in the springtime of the year. Through these grand rituals, powerful energies flowed into the world. Yet it was the renewal of the world or the sustaining of an abiding universe, not the irreversible and nonrepeatable original emergence of the world, that was taking place. Only an irreversible self-organizing world like the one we live in could provide this special mode of participation in the emergent creation itself. This irreversible sequence of transformations is taking shape through our own activities as well as through the activities of the multitude of the other component members of the universe community.

This is not a linear sequence, because the component elements of the universe move in pulsations, in sequences of integration-disintegration, in spiral or circular patterns, and, especially on Earth, in seasonal expressions of life renewal. On Earth, the basic tendencies of the universe seem to explode in an overwhelming display of geological, biological, and human modes of expression, from the tiniest particles of matter and their movement to the shaping and movements of the seas and continents, with the clash and sundering of tectonic plates, the immense hydrological cycles, the spinning of Earth on its axis, its orbiting of the sun, and the bursting forth of the millionfold variety of living forms.

Throughout this confused, disorderly, chaotic process, we witness enormous creativity. The quintessence of this great journey of the universe is the balance between equilibrium and disequilibrium. Although much of the disequilibrium fails to reach a new and greater integration, the only way to creativity is through the breakdown of existing unities. This is evident in the explosion of ancient supernovae, which released the atomic elements into the cosmos and, ultimately, our atmosphere. This "destruction" of stars eventually provided the basis for life on our planet. History also shows us that disturbed periods can be creative periods, for example, the early medieval period of

Europe and the period of breakdown in imperial order in China at the end of the Han period around the year 200 C.E.

So too in religion we see that great creativity is found in stressful moments. It was during a period of spiritual confusion that Buddha appeared to establish a new spiritual discipline. The prophets arose in the disastrous moments of Israel's life. Christianity established itself during the social and religious turmoil of the late Roman period. Now we find ourselves in a period of the greatest disturbance that the Earth has ever known, a period when survival of both the human and the natural worlds in their present modes of being is threatened. The identification of our human fate with the destiny of the planet was never more clear.

This new context of thinking also establishes a new context for liturgy. Presently, our liturgies give magnificent expression to the periods of seasonal renewal and also, at times, to significant historical events or personal achievements. Especially in these moments of renewal, in the springtime of the year, the psychic energies of the human community are renewed in their deepest sources by their participation in the profound changes within the natural world itself.

But now a new sequence of liturgical celebrations is needed. Even more than moments of seasonal renewal, these moments of cosmic transformation must be considered sacred. Only by a proper celebration of these moments can our human spiritual development take place in an integral manner, for these were the decisive moments in the shaping of both our human consciousness and our physical being.

First among these celebrations might be a celebration of the emergent moment of the universe itself. This was the beginning of religion just as it was the beginning of the world. The human mind and all its spiritual capacities began with this moment. As with originary moments generally, it is supremely sacred and carries within it the high destinies of the universe in its intellectual and spiritual capacities as well as its physical shaping and living expression.

Of special import was the rate of emergence of the universe and the curvature of space, whereby all things hold together. The rate of emergence in those first instants had to be precise to the trillionth of a second. Otherwise the universe would have exploded or collapsed. The rate of emergence was such that the consequent curvature of the universe was sufficiently closed to hold the universe together within its gravitational bonds yet open enough so that the creative process could continue through these billions of years and provide the guidance and energies we need as we continue to move into the future.

This bonding of the universe, whereby every reality of the universe attracts and is attracted to every other being in the universe, was the condition for the rise of human affection. It was the comprehensive expression of the divine love that pervades the universe in its every aspect and enables the creative processes of the universe to continue.

It might be appropriate to designate this beginning moment of the universe as the context for religious celebration and even for a special liturgy that could be available to all the peoples of the planet as they begin to sense their identity in terms of the evolutionary story of the universe. A list of other transformative moments might also be selected for celebration, since these moments establish both the spiritual and the physical contours for further development of the entire world. Among these supreme moments of transformation we might list the supernovae that took place as the first-generation stars collapsed into themselves in some trillions of degrees of heat, sufficient to bring the heavier elements into existence out of the original hydrogen and helium atoms, and then exploded into the stardust with which our own solar system and the planet Earth shaped themselves. This entire process can be considered a decisive spiritual moment and a decisive physical moment in the story of the universe. New levels of subjectivity came into being, new modalities of bonding, new possibilities for those inner spontaneities whereby the universe carries out its capacities for self-organization. Along with all this came the magnificent array of differentiated elements with the capacity for all the intricate associations that we now know. Indeed, Earth as we know it, in all its spiritual and physical aspects, became a possibility.

To ritualize this moment would cultivate the depth of appreciation for ourselves and for the entire creative process, which is needed just now, when the entire earthly process has become trivialized. Right now we have no established way of entering into the spiritual dimension of the story that the universe is telling about itself, the shaping of the Earth and of all living beings, and finally of ourselves.

The human is precisely that being in whom this total process reflects on and celebrates itself and its numinous origins in a special mode of conscious self-awareness. At our highest moments, we fulfill this role through the association of our liturgies with the supreme liturgy of the universe itself. Awareness that the universe is the primary liturgy has been recognized by the human community since the earliest times. The human personality and the various human communities have always sought to insert themselves into space and time through integration with the great movement of the heavens and the cycles of the seasons. What is needed now is integration with this new sequence of liturgies related to the irreversible transformation sequence whereby the world has come into being.

A great many of the mysteries of the Earth could be celebrated. We could go through the entire range of events whereby the universe took shape and inquire not simply into the physical reality but the religious meaning and direction of these events in their more comprehensive context. The development of photosynthesis is especially important, then the coming of the trees and later the coming of the flowers one hundred million years ago. Just how we would celebrate the birth of the human species is a challenge beyond all previous considerations.

Only such a sequence of religious celebrations could enable the cosmology of religions to come into being in any effective manner. If the sacred history of the biblical world is recounted with such reverence, so too should be the recounting of the sacred history of the universe and of the planet Earth. In all of this we can observe the continuity of the human religious process with the emergent process of the universe itself, with the shaping of the planet Earth, with the emergence of life, and the appearance of the human.

We find this difficult because we are not accustomed to think of ourselves as integral with or subject to the universe, to the planet Earth, or to the community of living beings. We think of ourselves as the primary referent and of the universe as participatory in our human achievements. Only the present threats to the well-being of the human as a species and to the life systems of the Earth are finally causing us to reconsider our situation.

This leads us to a final question in our consideration of the various religious traditions: the question of the religious role of the human as species. History is being made now in every aspect of the human endeavor, not simply within or between nations, ethnic groups, or cultures but between humans as species and the larger Earth community. We have been too concerned with ourselves as nations, ethnic groups, cultures, religions. We are presently in need of a species and interspecies orientation in law, economics, politics, education, medicine, religion, and whatever else concerns the human.

If until recently we could be unconcerned with the effect of human activities on the species level, this is no longer the situation. We need more than a national or international economy or even a global human economy; we need a species economy. A species economy will relate the human as species to the community of species on the planet, an economy that will be an integral Earth economy. Already this is beginning in the awareness that the human is overwhelming the entire productivity of the Earth. The human at the end of the twentieth century was using up some 40 percent of the entire productivity of the Earth. This has left an inadequate resource base for the larger community of life. The cycle is overburdened to such an extent that even the renewable life systems are being extinguished.

We could outline the need for a species, an interspecies, and even a planetary legal system as the only viable system in the present situation. We could say the same thing as regards medicine, since the issues of species health and the health of the planet are also intimately connected. Human health on a toxic planet is a contradiction. Yet we are, apparently, trying to achieve just that. The primary objective of the medical profession must be to foster the integral health of Earth itself. Only afterward can human health be adequately attended to.

In each of these cases—economics, law, medicine—the planet itself constitutes the normative reference. There already exists a planetary economics. The proper role of the human is to foster the economics of the Earth and to see that our human economies function in relation to and in service of the planetary economy. The same thing could be said for the realms of law and governance. There exists a comprehensive participatory governance of the planet. Every member of the Earth community rules and is ruled by the other members of the community in such a remarkable manner that the community as a whole and its individual members have prospered remarkably well over the millennia. The proper role for the human is to articulate its own governance within this planetary governance.

What is arising in human awareness is our nature as a species, which has emerged out of planetary processes. This awareness is beginning to reshape our religious imagination. This concept implies a prior sense of the religious dimension of the natural world. If the Earth is an economic mode of being as well as a biological mode of being, then it might not be too difficult to think of the Earth as having a religious mode of being. This seems to be explicit in many of the scriptures of the world, although this concept is yet to be articulated effectively in the context of our present understanding of the great story of the universe. In general, we think of the universe as joining in the religious expression of the human rather than the human joining in the religious expression of the universe. This has been the difficulty in most spheres of activity. We consistently think of the human as primary and the universe as derivative rather than thinking of the universe as primary and the human as derivative.

Our best model for this new vision within the context of a spatial mode of consciousness is probably found in the classical traditions of China. Within the perspective of a time-developmental mode of consciousness, models such as that developed by Pierre Teilhard de Chardin need to be further elaborated. It may be one of our greatest challenges to develop such an integrated cosmological perspective that celebrates the human as arising from and dependent on the universe.

CHAPTER 10

An Ecologically Sensitive Spirituality

I REMEMBER being in Italy in Umbria, on the western slope of the Apennines, bathed in the soft summer light of this region, just as Giotto (c. 1267–1337) and the Umbrian school of painters must have experienced it. What we see here now is only a remnant of the scene enjoyed by St. Francis (1181–1226) and his early companions. The quiet lanes have been replaced by paved roads; the donkey-drawn carts have been replaced by automobiles. We feel an intimacy with these earlier times. But we also breathe an atmosphere less refreshing; the acrid taste of automobile fumes saturates the air. A crowded world has emerged on the scene. The beginning of the modern commercial world that St. Francis perceived with a certain foreboding in the opening years of the thirteenth century has developed into the industrial centers of the late twentieth century. The consequent assault on the natural world is leading to a certain anxiety concerning the future course of human affairs.

We should spend some time in thoughtful brooding over the decisions we need to make at present for the well-being of the Earth community. To understand the challenges of our times, we might go back to the opening years of the thirteenth century, the period of St. Francis of Assisi, when our present world began to take shape. This time of high spiritual accomplishment in the European world was when the commercial spirit entered more robustly into the Western soul. The cities of Europe were reborn, after a long decline following the dissolution of Roman order in the fifth century. The Hanseatic League of commercial cities in northern Europe was formed in the thirteenth century. Venice had begun its commercial empire somewhat earlier, during the crusades from 1095 until 1291. This intensive commercial activity culminated in the fifteenth- and sixteenth-century overseas ventures of the seacoast peoples of Europe, leading to the discovery of America. With this discovery, the European quest for dominance of the entire planet was begun. It seems appropriate, then, to speak of the European occupation of North America as one of the most momentous periods in world history.

The historical role of North America is something of a parable of the larger human process, for when that first tiny mast of a European ship appeared over the Atlantic horizon the indigenous peoples could not have imagined the domination that would follow. The peoples from across the sea might have come to join the great community of life on this distant continent. They might have responded to the spiritual grandeur of the forests, the rivers and the woodland creatures, and to the mountains and valleys with reverence and wonder. They might have learned from the native peoples something of the spirituality integral to this land.

Unfortunately, these people from across the sea thought they already knew everything. They brought with them a book, the Bible, as their primary reference as regards reality and value. Though a work of great spiritual significance, this book has also been used to justify the domination of peoples and land in various parts of the world. Moreover, the book has contributed to the inability of humans to see the natural world as revelatory. Revelation was in scripture alone, not in nature itself.

The North American continent was ready to offer a profound spirituality to the incoming peoples. In the magnificence of its natural splendor, in the grandeur of its forests, in the beauty of its rivers, in the abundance and variety of its wildlife, this continent still had much of its primordial vigor, something of the innocence that older civilizations had lost long ago. In all these ways it was a more immediate manifestation of the divine than the incoming peoples had experienced for centuries.

Yet to have responded to this pervasive presence of the natural world would have been considered inappropriate by a people accustomed to experience the spiritual order of things in terms of the biblical world. We might consider this inability to enter into any significant rapport with the primordial world as one of the sources of our present problems of spirituality and sustainability. Because the spiritual dimension of this continent could not be recognized or responded to in any adequate manner, no proper reverence was given to the continent so as to mitigate the exploitation of the immense wealth available here.

This alienation was further strengthened by the humanist formation of Western civilization, which fostered the exaltation of the human over the natural world. Both the spiritual and the humanist dimensions of the Western tradition had only minimal concern for the natural world. Education that should be oriented toward deepening the intimacy of the human inhabitants with the larger Earth community and the comprehensive universe community was turned away from the outer world and driven inward toward self-appreciation of the human and consequent exploitation of the nonhuman.

This attitude was further strengthened by the Newtonian cosmology set forth in the seventeenth century. After the material explanation of the universe given there, the natural world could no longer carry the same spiritual significance. This was now a world of objects to be manipulated for the benefit of the human. Already in the first part of the seventeenth century, René Descartes (1596–1650) had laid the groundwork for an assault on the planet by his division of the universe into mind and matter. What was not mind was mechanism.

With this background it is little wonder that when the incoming peoples arrived in America they had no deep feeling for the natural world and none of the aesthetic appreciation shown in earlier times. Above all, they had no awareness that humans form a single integral community with the other components of the continent, with the planet Earth, and ultimately with the universe. The nonhuman world was seen as a collection of objects to be exploited, not as subjects to be communed with. We have continued this exploitation in these past four centuries with such a passion that the devastation has flowed over into the larger dimensions of the planet, and now we are at a planetwide impasse as regards human consumption and Earth's limits. These two are on a collision course.

While Earth's resources are finite, what is not limited is our desire to understand, to appreciate, and to celebrate the Earth. We do need endless progress, but not, however, in material development. Only an increase in aesthetic appreciation and spiritual experience can be without limit. Advance in material possession and use is severely limited.

Our most urgent need at the present time is for a reorientation of the human venture toward an intimate experience of the world around us. If we would go back to our primary experience of any natural phenomena—on seeing the stars scattered across the heavens at night, on looking out over the ocean at dawn, on seeing the colors of the oaks and maples and poplars in autumn, on hearing a mockingbird sing in the evening, or breathing the fragrance of the honeysuckle while journeying through a southern lowland—we would recognize that our immediate response to any of these experiences is a moment akin to ecstasy. There is wonder and reverence and inner fulfillment in some overwhelming mystery. We experience a vast new dimension to our own existence.

Our rediscovery of the mystique of Earth is a primary requirement if we are ever to establish a viable rapport between humans and the Earth community. Only in this context will we overcome the arrogance that sets us apart from all other components of the planet and establishes a mood of conquest rather than of admiration. To assume that conquest and use are our primary relations with the natural world is ultimate disaster not only for ourselves but also for the multitude of other living forms on the planet.

To lessen the grandeur of the outer world is to limit the fulfillment available to our inner world. For the stars in the night sky over our cities to be blocked from view by particle and light pollution is not simply the loss of a passing visual experience, it is a loss of soul. This is especially a loss for children, for it is from the stars, the planets, and the moon in the heavens as well as from the flowers, birds, forests, and woodland creatures of Earth that some of their most profound inner experiences originate. To devastate any aspect of the natural world is to distort the sublime experiences that provide fulfillment to the human mode of being.

We need to move from a spirituality of alienation from the natural world to a spirituality of intimacy with the natural world, from a spirituality of the divine as revealed in the written scriptures to a spirituality of the divine as revealed in the visible world about us, from a spirituality concerned with justice only for humans to a spirituality of justice for the devastated Earth community, from the spirituality of the prophet to the spirituality of the shaman. The sacred community must now be considered the integral community of the entire universe and, more immediately, the integral community of the planet Earth.

Our Western Christian humanist world needs to experience a reversal of values. We live in a time when the survival of humans can only be achieved by saving the natural world upon which humans depend for both their psychic and physical flourishing. While we have already outlined the basic psychic need we have for the natural world, we should also mention our need for water, air, nourishment, shelter, and a sense of security in the presence of the grand complex of living and nonliving forces that make up the integral community of Earth.

A pervasive flaw in Western civilization is the attitude that only the human is capable of having rights. The attitude that the primary purpose of the nonhuman world is its use by humans can no longer be accepted. This attitude has contributed to our devastation of the natural world. In reality, every being has three basic rights: the right to be, the right to habitat, and the right to fulfill its role in the great community of existence. Likewise, every being has a right not to be abused by humans, a right not to be despoiled of its primary dignity whereby it gives some manner of expression to the great mystery of existence, and a right not to be used for trivial purposes.

To bring about a recognition of this new sense of the human role in relation to the natural world, we need a radical transformation throughout the entire human venture. The dynamics of the industrial-commercial-financial empires of these times is driving the Earth into a termination of the Cenozoic Era in the geobiological story of the planet. This period, the Cenozoic, the last sixty-five million years, has been the culmination of the most brilliant phase of life's expansion on the planet. Only at the end of this period, when the planet was at its most gorgeous expression, was it possible for humans to appear. For only in a world of such magnificence could the human mode of being be fully developed, only then could the divine be properly manifested, only in such a world could the burden of human sensitivity and responsibility be sustained, the human condition be endured, and the constant healing needed by the human soul be effected.

This magnificence was not recognized by the settlers of the North American continent nor by their heirs, who later spread the nature-exploiting industrial enterprise around the planet. During this time, the spiritual and intellectual guides of our Western tradition have shown themselves to be inadequate to their task, however adequate they may have been in former times. A new type of spiritual guide is needed. Previously, Benedictine monks established themselves as the guides for our Western endeavor, by cultivating the soil through physical labor and by copying and explaining the great literary works of the past through their intellectual effort. Later in the medieval period, when the cities of Europe were reestablished, it was the new spirituality of the cathedral builders, university professors, and mendicant friars who guided the course of human affairs. In the medieval period came the political, social, and economic establishments that brought about a sense of national identity and later a sense of the people as competent to determine their own destiny.

Toward the end of the nineteenth century, a new set of guides appeared: research scientists, technologists, engineers, and above all corporate leaders. These persons were determined, through technological exploitation of the planet and its resources, to lead humans into a new golden age. The corporations, supported by the dominant political forces, were determined to seize power over every aspect of the planet. This effort at control has led to our present impasse in human-Earth relations. In particular, it has led to the radical disruption of the major life systems of the planet.

Throughout this modern period, the traditional spiritual leaders—scholars, religious teachers, and social reformers—have been unable to provide sufficient guidance. They have failed to recognize that the basic issue is not simply divine-human or interhuman relations but human- Earth relations and, beyond that, relations with the comprehensive community of the entire universe, the ultimate sacred community. This failure has led to the plundering of the planet by good persons, even deeply religious persons, for the supposed temporal and spiritual benefit of the human. This plundering of the planet to serve human purposes is what needs to change. The industrial movement, with its ideal of subjection of the planet, must now give way to the ecological movement. Only such an ideal will sustain the integral functioning of both the human and nonhuman components of the planet in a single integral community.

This ideal requires a new spirituality. We need the guidance of the prophet, the priest, the saint, the yogi, the Buddhist monk, the Chinese sage, the Greek philosopher, and the modern scientist. Each of these personalities and their teachings are immensely important in their own proper field of functioning. Yet, for these times they might all be considered limited as guides to the human process in its rapport with the natural life systems of the planet. We now have a new understanding of the universe, how it came into being and the sequence of transformations through which it has passed. This new story of the universe is now needed as our sacred story. Few of the traditional spiritual guides seem able to accept this understanding as a revelatory experience. This can only be done by an ecologically sensitive personality.

We need an ecological spirituality with an integral ecologist as spiritual guide. While we can expect this to be realized in only a partial and inadequate manner in any individual, we can still assert that such a spiritual personality is needed. We can also say that the spiritual ideal of former ages was realized in an unlimited variety of individual personalities and rarely in a manner sufficiently striking to become a referent for imitation by others. So too will the ecologist serve as a guide for these times. The great spiritual mission of the present is to renew all the traditional religious-spiritual traditions in the context of the integral functioning of the biosystems of the planet. This is what the project that began at Harvard at the Center for the Study of World Religions has undertaken. With a series of ten conferences, books, and a Web site, over eight hundred religious scholars, scientists, and activists have examined the resources of the world's religious traditions to meet the spiritual and ethical challenge of the environmental crisis. Ten volumes have been published, a journal has been established, and a Forum on Religion and Ecology is now located at Yale.[1]

Until recently, there has been a feeling in most religious traditions that spiritual persons were not concerned with any detailed understanding of the biological order of Earth. Often, the spiritual person was in some manner abstracted from concern with the physical order of reality in favor of the interior life of the soul. If attention was given to the physical order, this was generally in the service of the inner world. This neglect of attention to the natural world permitted those concerned with the more material things of life to take possession of the planet's land and wealth. It permitted the exploitation of the natural world for human gain. The integral ecologist can now be considered a normative guide for our times. The integral ecologist would understand the numinous aspect of a universe emergent from the beginning. The sequence of transformative moments of the universe would be understood as cosmological moments of grace to be celebrated religiously with special rituals. But above all, these moments would appear as revelatory of the ultimate mystery of the universe itself.

The integral ecologist is the spokesperson for the planet in both its numinous and its physical meaning, just as the prophet was the spokesperson for the deity, the yogi for the interior spirit, the saint for the Christian faith. In the integral ecologist, our scientific understanding of the universe becomes a wisdom tradition. We will finally appreciate that our new understanding of a universe that comes into being through a sequence of irreversible transformations has a revelatory dimension. This fresh understanding of the universe establishes a horizon under which all the traditions will henceforth need to function in their integral mode of self-understanding.

This issue of our human disturbance of the most basic life systems of the planet Earth is such that from here on, for an indefinite period, the main difference between human beings will not be the difference of conservative or liberal, based on political, social, or cultural orientation, as has been the case for humans in the Western world throughout the twentieth century. Rather, it will be the difference between the entrepreneur and the ecologist, the difference between those who exploit the planet in a deleterious manner and those who sustain the planet in its integral functioning. This difference will provide not only the public identity of individuals; it will also be a primary designation in the professions: law, medicine, education, religion, or politics. The prefix "eco-" will occur in a multitude of words that will refer to the coherence of ideas, actions, or institutions in relation to the integral life systems of the planet.

The seriousness of the situation we are discussing can hardly be exaggerated. It is the issue of life and death, not simply for human individuals or the human community. Rather, it is the issue of survival of the most gorgeous expression that the ultimate forces of the universe have given of themselves, so far as we know. In designing a program that can adequately deal with these issues, we need to be concerned with principles, strategies, and tactics. Tactics, for example, involves numerous actions such as recycling materials, limiting our use of energy, composting, conserving water supplies, insulating our buildings, and a multitude of adaptations of a similar nature.

Key strategies include education, namely teaching children about the natural world, how its living systems function, and how humans fit into these systems. It would involve dealing with corporations so as to assure control over emissions from industrial production. It would necessitate working with city-planning boards to determine land use in a given territory. One of the most significant strategies would be interaction with the universities. These educational institutions need to understand that ecology is not a course nor a program. Rather, it is the foundation of all courses, all programs, and all professions, because ecology is a functional cosmology and the universe or the cosmos is the only self-referent mode of being in the phenomenal world. Every other being is universe referent. Cosmology, or the universe story, is the implicit basis of every particular course or program.

Beyond these considerations is the question of principles, guiding rules governing the course of human action. We are involved in a deep cultural pathology, demanding a cure. What is most needed in addition to the new technologies integrating our human needs with solar energy and the organic functioning of planetary life systems is a deep cultural therapy that will identify the sources of our pathology and provide a way of returning to the jubilant life expression that should characterize any human mode of being.

I propose that one of the most fundamental sources of our pathology is our adherence to a discontinuity between the nonhuman and the human, which gives all the inherent values and all the controlling rights to the human. The only inherent value recognized in the nonhuman is its utility to the human. This discontinuity between the human and the nonhuman breaks the covenant of the universe, the covenant whereby every being exists and has its value in relation to the great universe community. Nothing bestows existence on itself. Nothing survives by itself. Nothing is fulfilled in itself. Nothing has existence or meaning or fulfillment except in union with the larger community of existence.

In the phenomenal world, only the universe is self-referent. Every being in the universe is universe referent. Only the universe is a text without a natural context. Every particular being has the universe for context. To challenge this basic principle by trying to establish the human as self-referent and other beings as human referent in their primary value subverts the most basic principle of the universe. Once we accept that we exist as an integral member of this larger community of existence, we can begin to act in a more appropriate human way. We might even enter once again into that great celebration, the universe itself.

PART IV

CHAPTER 11

The Universe as Divine Manifestation

THIS CONTINENT is the immediate context of our lives. I speak primarily about North America, not about the planet or about nature or the world or creation. Perhaps we could speak about the Mississippi River, the Delta region, the swamp cypress, the bayous, the marine life that inhabits this area, the oaks, the pines. We might also speak about the wonders of this region of the continent, such as the birds that inhabit it.

The wondrous moments of our lives should be more frequent than they are in the civilization that we have contrived for ourselves. The comforts of our lives have diminished the wonder. Not only do we miss the dance of life on the planet, but we also fail to see this dance in the universe in which our planet Earth floats—the sun, the stars of the zodiac, the Milky Way galaxy.

The religious ceremonies of many peoples of the world were associated with the various celestial phenomena. In China, the coordination of the entire range of human activities was expressed according to the celestial configurations and the sequence of the seasons. The period when the warmth returned and all that had gone into the realm of death returned to life was given the name of spring. Spring means to "burst forth"—just as we burst forth into laughter or into tears when our inner self can no longer be contained.

The cosmological context of life is still with us. The day-night sequence determines when we live our lives and carry out those functions necessary for survival. Planting and harvesting occur in the proper season. When we write a letter, we indicate the day and the year and the place where this is taking place. We are here just now because the sun has moved below the horizon and the mystery of night has enfolded us. This has always been the moment when humans gather around the fireplace and tell their stories. This is the mystical moment when the Earth has completed its rotation around the Sun. At this moment of cosmological change, we pause to consider the basic elements of existence. We reintegrate ourselves with the universe around us, the universe that has to some extent been distanced during the day and now must reintegrate itself. Living beings are exhausted in their physical energy and need to restore themselves. The wonder of the night brings the quietness, the calm, the healing of the fever of the day.

We are placed in the universe by the naming of the days of the week and the months of the year. This cosmological orientation is a linking of time to the planets and the solar system. These days of the week are thus qualitatively different, each day having its own mystique or spirit power associated with it. The seven-day week was invented during the Babylonian period and then adopted by the Greeks and Romans. The days were designated in this way:

Sunday for the Sun
Monday for the Moon
Tuesday for Mars: Tiu or Tyr, Teutonic god of War
Wednesday for Mercury: Wodin
Thursday for Thor: Jove
Friday for Venus: Frigg, a wife of Odin
Saturday for Saturn: Sabbath

The months of the year also have cosmological or historical associations: January for Janus, the Roman god identified with doors and beginnings; February is for Febbrua, a Roman feast of purification; March, for Mars, the Roman god of war; April, from a Roman reference to Venus; May, from the Roman goddess Maia; June, from Juno, wife of Jupiter, the queen of heaven and goddess of light; July for Julius Caesar; August from Caesar Augustus. The closing months of the year are numbered: September (seventh), October (eighth), November (ninth), and December (tenth), since the year began with March.

Morning is the beginning of the day. It is named "sunrise," as we have not yet changed our thinking and our language to more accurately reflect the turning of Earth toward the sun, beginning a new day-night cycle. Morning is beginning, energy, wakefulness, the period of work. Night is the ending period and the quieting of activity, when the visible world is dimmed and shadows fall over half of the planet. It is the time when fireflies appear and signal with their momentary flashes of light. Night is also associated with dreams and deepened consciousness. However, nocturnal creatures come out and keep the pulsations of life continuing without interruption.

The story of the Western world is the story of how the peoples whose culture took shape through the religious inspiration of the Hebrew Bible, the New Testament, the humanism of the Greek world, the political-legal genius of the Romans, and a brilliant medieval period became so entranced with a secular, scientific, industrial civilization serving limited human needs that it was willing to devastate the entire planet for the immediate benefits received. Their assault upon the Earth has been so violent in modern times, both to its geological structures and its living species, that we face a tremendous crisis.

We see this assault especially on the North American continent. In just a few centuries of occupation, the Earth has become extensively ruined in its forests, desolate in its loss of animal species, diminished in the fertility of its soil, toxic in its atmosphere, polluted in its rivers, diminished in the marine life that once flourished abundantly around its shores, and threatened by radioactive waste without means of disposal.

How all this could happen remains unclear, although it seems to have been caused by our exploding population, our technological power, and our entrancement with creating the industrial world. What is especially difficult to explain is that this destructive relationship with the planet Earth has become so institutionalized, so integrated into the cultural structure and functioning of the planet, so integral with its legal principles, and so built into the educational life from kindergarten to university and professional education that it has become a culture of technology with little reference to natural systems. This new situation is so inherent in cultural structures that it seems to be the inevitable consequence of the religious, political, educational, ethical, and economic establishments of Western peoples.

We are aware of the difficulties that have occurred in human relations with Earth. But we are especially concerned that the relationship has become fixated, pervasive, and resistant to remedy. This fixation can be traced back to these very guiding elements. The situation has been so easily accepted by the religious establishment that scarcely any protest is heard from that source. Moreover, the legal establishment is so much a part of the commercial venture that law schools teach the principles that allow these violations of the planet. That the history of Western civilization should still be taught as the way to fulfillment of the human personality, that the human should still assume that the greater the exploitation of the natural world the more fulfillment will be experienced by its human population, that in these early years of the twenty-first century the Congress of the United States should dismiss even the minor achievements toward rectifying this situation—these are all cause for concern.

That schools of management should still be fostering more extensive exploitation of the natural systems of the planet is also cause for reflection. That biologists should still be fostering disruption of the natural life systems through genetic engineering in the conviction that they can remedy the disorders of the present living forms on the planet is disturbing. That the rights of natural modes of being are not recognized by human establishments on a global scale is yet another reason to be concerned.

Scientists are in immediate contact with the phenomena of the natural world. Astronomers observe the stars so as to learn their physical measurements for the benefit of humans. They congratulate themselves that they are able to plot so much of the natural phenomena through their telescopes, but they enter into such limited communion that they often miss the possibility for dynamic relationship.

Analogy is the key to all human communion with the nonhuman, whether the divine or the natural world. The divine has ways of speaking that are not human ways. So too do natural phenomena have ways of speaking that are not human language. The effort to reduce all wisdom to a univocal language is a primary error or failure of our times. To think that the various natural phenomena, such as stars, do not speak to us is to break with natural systems. So, too, to think that the divine does not speak to us is also an error. In early times, this break or separation between human language and the language of other natural phenomena was not evident. This sense of human/nonhuman language goes back to the fact that the divine communicates to us primarily through the languages of the natural world. Not to hear the natural world is not to hear the divine.

Part of the difficulty of not hearing the language of the natural world is that we have limited our understanding of speech to persons, not nature itself. Persons speak; nature does not, except to the poets. A "person" is defined as a living human being as opposed to a nonhuman animal or an inanimate thing. A sense of personhood as a distinct mode of being is identified with having a human mode of consciousness and using language to communicate.

Human language arose, however, not only as a means of attunement between persons but also between humans and the natural world. The belief that meaningful speech is a purely human property was entirely alien to those oral communities who first evolved our various ways of speaking. By holding to such a belief today, we may well be inhibiting the basic function of language. By denying that birds and other animals have their own modes of communication, by insisting that the river has no real voice and that the ground itself is mute, we stifle our direct experience. We cut ourselves off from the deep meanings in many of our words, severing our own language from that which supports and sustains it. We then wonder why we are often unable to communicate even among ourselves. We need the poets and artists to restore this forgotten language.

Ecologically, it is not primarily our verbal statements that are "true" or "false" but rather the kind of relations that we sustain with the rest of nature. A human community that lives in a mutually beneficial relation with the surrounding Earth is a community, we might say, that lives in truth. The ways of speaking common to that community—the claims and beliefs that enable such reciprocity to perpetuate itself—are, in this important sense, true. They are in accord with the right relation between these people and their world. Meanwhile, statements and beliefs that foster violence toward the lands, ways of speaking that enable the impairment or ruin of the surrounding field of beings, can be described as false, because they encourage an unsustainable relation with the encompassing Earth.[1] A civilization that relentlessly destroys the living land it inhabits is not well acquainted with that land, regardless of how many supposed facts it has amassed regarding its calculable properties.

We need to establish a rapport among the divine, the natural, and the human. These three each have their proper language. We need to understand that the locus of the meeting of the human and the divine is in the natural world. The voice of the natural world is the resonance of the divine voice. Here the human enters into the divine order, since the divine in itself is not directly accessible to human intelligence or understanding. The human in its own structure and functioning is also a manifestation of the divine. But an inner activation of the divine is not possible by humans alone. We need the outer world to activate the inner world of the human.

I have often said that the wonder and beauty of the natural world is the only way in which we can save ourselves. Just now we are losing our world of meaning through our destruction of the natural world wherein the divine speaks to us. The more we are absorbed into our own selves, the less competent we become in our patterns of communication with the outer world. So too the more shriveled we are in our inner world.

The biblical prohibition against communication with false gods has become a way of dissolving our relationships with the powers of the nonhuman world about us. It has lessened our ability to enter into that world, where the divine is manifested in such meaningful ways.

Anywhere on Earth we awaken to ourselves in the midst of a remarkable setting, whether in the luxuriant forests of the Amazon, on Mount Kilimanjaro in Africa, in the jungles of Indonesia, or the shores of Lake Baikal in Siberia. In New Orleans, where the great central valley of the North American continent turns into its delta region, there is the river, the bayous, the flow of the warm air, the variety of vegetation, the wind in our faces, the sight and hearing and feeling of the warmth of the day and the chill of the evening, all of this bringing us into an intimacy with an outer world. The only way of talking about the universe is to speak of the immediacies about us.

Although our surroundings are intimate, we seem constantly to be defending ourselves against them. We speak of conquering the natural world. In the great urban centers of our modern world, we do not initiate children into the mysteries of the bioregion about them; instead we put them into a classroom and insist that they become literate, to read what humans have written. We speak of the Western tradition, the religious traditions, cultural traditions, political traditions, scientific traditions. Human traditions are everything.

Even though we depend on traditional learning for any integral interpretation of experience, we also need the immediacy of experience. The difficulty is in being alienated from primordial experience. To have the interpretation without the experience is the present difficulty. We are alienated from immediacy with the surrounding natural community to which we belong and which is constantly communicating with us. Because we live in a human-made environment, the challenge is how to keep this immediacy with the natural world and to establish a traditional wisdom that deepens our understanding of the experience.

Religion takes its origin here in the deep mystery of what we see, hear, touch, taste, and savor. The more a person thinks of the infinite number of interrelated activities taking place throughout the natural world, the more mysterious it all becomes—the more meaning a person finds in the May blooming of the lilies, the more awestruck a person might be in simply looking out over some little patch of meadowland. While we sometimes long for the overwhelming experience of the western mountains, the immensity or the power of the oceans, or even the harsh magnificence of desert country, we can also relish the tiny stream flowing beneath overhanging willow branches or the sight of the sky at sunset.

Such experiences were more available before we entered into an industrial way of life, with, for example, our electric lights, which do not permit us to experience the night in the depths of its mystery or the starry heavens in any reflective context. In earlier times, the universe, as manifestation of some primordial grandeur, was recognized as the ultimate referent in any human understanding of the magnificent yet fearsome world about us. Every being achieved its full identity by its alignment with the universe itself. Indigenous peoples of the North American continent situated every formal activity in relation to the six directions of the universe, the four cardinal directions combined with the heavens above and the Earth below. Only thus could any human activity be fully validated.

Usually I speak about a wonder-filled intimacy with planet Earth, with the sun, the moon, the stars, and eventually with the universe entire. Here I am focusing on our presence on the North American continent. For the most part, we who reside on this continent have come from other parts of the planet, from Europe, Africa, and Asia.

As humans we function differently from other living species, which are determined in their life patterns and in their association among themselves and with other species and have much less of that psychic development we identify as human consciousness. The genetic coding of the nonhuman species establishes their inner patterns of consciousness, their capacities for acquiring their food, their patterns of mating, their social structures, and their communication methods. Some species do have extensive acculturation processes that constitute learned behavior: bears need teaching as regards their capacity for fishing and insects such as bees carry out remarkable processes to produce honey.

What we have learned during these past few decades concerning the special insights, the functional skills, and the modes of consciousness of various animal species is exceptional. I speak especially of the work of Jane Goodall among the primates, particularly the chimpanzees. From her work and from that of others, we are finding that the power of reciprocal communication in the animal world is far greater than we had previously thought.

Scientific studies of the universe have given us amazing information on the structure and emergence of the universe. Although we increasingly know more about the universe and its evolutionary processes, we have less intimacy with it. We do not celebrate the universe. It has lost for us its mystical dimensions. We live in a less meaningful world than those who preceded us—certainly less meaningful than other civilizations that have had far less information about the universe.

Indeed, one of the most astonishing statements about the universe is Steven Weinberg's reflection at the conclusion of his book *The First Three Minutes: A Modern View of the Origin of the Universe*. There he tells us that the more we know about the universe the more pointless it seems to be. How different this is from the poet William Blake (1757–1827), who asked: What do you see when you look out over the landscape? Do you simply see the sun rising or do you see the flaming forth of the deep mystery of the universe?

I remember attending a Global Forum in Oxford in 1988, where some two hundred scientists, religious thinkers, and political representatives gathered to discuss the future of the planet and its relation to the human community. Behind the speakers' podium was an enlarged reproduction of the photograph of the planet Earth as seen from space—a blue-and-white globe majestically sailing through the dark. Yet one of the most prominent members of the symposium remarked to me with a certain concern that this was not the planet Earth in any meaningful way. That is all he said. I was somewhat puzzled at first by what he could possibly mean. Then it occurred to me that it was the very physical splendor of Earth as presented that he somehow found inadequate. It did not present the soul of the planet. It did not show the grasses, flowers, or meadows of the planet; it showed no deserts, rainforests, rivers, lakes, or vegetation. There were no trees, no soaring birds or butterflies, and no animals moving about on the plains or through the woodlands. Instead it was a colorful marble hung in the sky, a small sphere such as we used to play with in childhood games. This photo showed Earth in such an entrancing way that it distracted from the more particular aspects of the planet and thus of the further implications of the nature and structure and functioning of the universe itself.

It is truly astounding that we have such insight into the functioning of the universe, that we know Earth and its biosystems and the mysteries of genetic coding, that we can manipulate Earth and biological organisms so extensively, that we can deal with electronics and microengineering at the atomic level, and that we can set up such amazing communications programs. Yet in all of this there is something that eludes us. There is something completely out of proportion, since our knowledge has not led to an expansion of our emotional feeling, our aesthetic appreciation, or our sense of the sacred. Nor has it increased our wonder.

We have come to know so much about God from our scriptures and our theological and religious traditions that somehow we have lost our sense of wonder. It seemed that we had control of God. God became reduced to our ideas of God, and belief in God became a sterile commitment on our part. Some exceptions are, of course, the novels of Fyodor Dostoyevsky (1821–1881), the romantic poets, and the poetry of T. S. Eliot (1888–1965). There is a sense of the sacred in relation to the universe in the naturalist writers, especially with John Muir (1838–1914) and with Henry David Thoreau (1817–1862), and with many nature writers in the present period. Rachael Carson (1907–1964) roused us to a sense of the mystery of things in her book *A Sense of Wonder*. While it is true that the music of Beethoven, Bach, Mozart, Handel, Haydn, and Bruckner carried intense religious expression, this music was disassociated from the formal religious life of the society. It was largely relegated to the concert hall rather than the chapel.

Wonder is that which arouses awe, astonishment, surprise, or admiration: a marvel, a feeling of glory. Glory is described by Saint Thomas as *clara notizia cum laude*: clear knowledge with praise; to express strong approval or admiration for; to applaud, extol, commend; to exalt. This is the great challenge of the human at present—to recover the language of wonder and praise. Then we can give expression to the deep reciprocity and relatedness at the heart of the universe. In this way we may take up the immense challenge of restoring our world.

CHAPTER 12

The Sacred Universe

TO UNDERSTAND America, we need to reflect on earlier times, when the human community was experienced within a universe of subjects to be communed with, not of objects to be exploited. Humans were intimate members of this single community of life. Everything, from the stars in the heavens to the flowers, insects, animals, and humans of Earth, constituted a comprehensive sacred community. Into this community we were born, nourished, educated, guided, healed, and fulfilled. The surrounding powers brought us into being, sustained us, and led us to our destiny.

It was a world alternating between scarcity and abundance, between the heat of summer and the chill of winter. Yet it was a livable world with deep intellectual, artistic, and emotional fulfillment, especially in rituals celebrating the seasons of the year: in spring the exuberance of life, in autumn thanksgiving for the abundance of sustaining gifts. Every aspect of daily life had significance. A gorgeous sense of bodily ornament existed. The entire natural world was interpreted with extravagant symbolism. Everything had its special identity. The four cardinal directions each had unique mystical significance. Even when the various archaic and classical civilizations developed, this sense of living in a world of mysterious powers continued.

Consciousness of such a rapport with the universe began to fade in the postmedieval period, as the world flattened into the vision of René Descartes (1596–1650). The mythic intimacy we had with the spirit powers of the natural world began to diminish and then disappear. We withdrew from any feeling of identity with the natural world. The world around us became a natural resource to be used, not a vital reality to be communed with. As their use value increased, the inherent wonder and meaning of things decreased. As we diminished other modes of being, we exalted ourselves. We were superior beings, masters of the world, destined to redesign the planet for human benefit. For example, in the early seventeenth century Francis Bacon (1561–1626) suggested that we put nature "on the rack," to force her to yield up her secrets so that we could control the entire natural world.

This exaltation of the human reached a new level of expression when Europeans began to occupy the North American continent. These peoples were escaping from the oppressions of European monarchies and their feudal servitudes to create a new mode of human community. To paraphrase Thomas Paine (1737–1809), never since the time of Noah had there been such an opportunity to remake the world.[1] The Europeans arrived to take charge of the continent and its future destiny. They did not recognize that this continent had its own spirit presences and its own laws that they should obey if they wished to live here. These new settlers might have had a certain reverence for the world they were entering. They might have had some care for how they conducted ourselves. The pervasive brashness in the conduct of the Europeans in this newly discovered land is noteworthy.

In the late fifteenth century, Christopher Columbus (1451–1506) found the native peoples of the Caribbean islands so sweetly tempered and so gracious and helpful to the strangers in their midst that he wondered whether they had suffered the effects of any "original sin." Nevertheless, he captured a number of the islanders and took them as slaves back with him to Spain. In the early seventeenth century, Arthur Barlowe (1550–1620), an English writer from the Virginia colonies, tells us that "a more kind and loving people there cannot be found in the world."[2] Faced with the aggressive attitude of the colonists, Wahunsenacawh (1547–1618), an exasperated chief of the Powhatan, confronted John Smith (1580–1631) with the questions: "Why do you take by force what you may have quietly by love? Why will you destroy us who supply you with food? What can you get by war? . . . We are unarmed and willing to give you what you ask if you come in a friendly manner and not with swords and guns, as if to make war upon an enemy."[3] John Lawson (c. 1674–1711), an early explorer of the Carolina region, tells us that as he walked through the region from Wilmington northwest to the Guilford area and then northeast to New Bern, he found the Indians more generous to the settlers than the settlers were to the Indians.

What is said of this generosity of the Indians can be said of the natural world and its willingness to give us an abundance of what we need if only we realize that at the human level we must impose limits upon ourselves. Other modes of being have opposed modes of existence, or limiting conditions, so that no single one dominates or completely overwhelms the others. Unfortunately, we are a conquering people with a deep anxiety in approaching other peoples or modes of being. We only knew how to deal with the world beyond ourselves by antagonism and conquest. We even invented the term "Manifest Destiny," as though some heavenly commission had conferred on us the right to unlimited conquest over this continent and beyond.

This self-exaltation found expression in the U.S. Constitution, written in 1789, a constitution that was the fulfillment of the anthropocentric ideals of Western civilization and at the same time an undermining moment of Western civilization. It was the fulfilling moment because it gave humans the highest status they had ever attained in any political establishment known to history. It was an undermining moment because the Constitution gave humans limitless rights to possess and use this continent, while giving no rights to the continent and no protections whatsoever against its human predators.

According to the Constitution, government was to protect humans, not the integral community of the land and its people. Government was not to intrude on personal and private human affairs, nor was government to interfere with citizens' acquisition, possession, use, or disposal of property. Even with government-owned property, private citizens had rights to mine the land for a minimal fee and to possess, free from interference, any minerals extracted. So, too, with grazing rights, logging rights, and irrigation rights—these were all human rights. Government was to protect the absolute freedom of citizens to dispose as they pleased of this new land. The continent became vulnerable to the violation of its most sublime expression.

In fulfillment of this role to occupy and use the resources of this continent, we have dedicated our educational programs primarily to the pursuit of so-called useful knowledge, not to knowledge as intimate presence and participation in the wonder and magnificence of the universe, through which we find the meaning of our existence. Use has acquired something of a sacred role in the American mind. What is not used is considered wasted.

This utilitarian disposition had been kept within limits throughout the earlier periods of Western civilization. Then, in the early seventeenth century Descartes (1596–1650) articulated a mode of consciousness based on the conviction that the natural world is devoid of any spiritual principle in either its plant life or its animal life. With this thinking, the natural world was stripped of its ensouled qualities and became a deposit of natural resources. The empirical sciences, expressed with mathematical precision, became the only valid ways of knowing.

At first, our capacity to exploit the continent was restrained. We could do only limited damage. But gradually, through the untiring efforts of an immense number of research scientists, we came to greater insight into the nature and functioning of the natural world and the means whereby we might exploit the continent for our own convenience and commercial profit. We began our relentless experiments to find ways of using for human benefit alone this new land.

Such research involved precise observation, hypothesis, experimentation, measurement, and interpretation. These processes also determined how we would use our minds. An analytical understanding was required. Only poets were permitted to commune with power presences throughout the cosmos. Other intelligent humans were occupied with those realms of understanding whereby we could exploit the energies flowing all around us. The wonder of Earth diminished as its utilitarian value increased.

We logged the forests, mined the hills, dammed the rivers, and shot to extinction billions of passenger pigeons, the most numerous bird species ever known to exist. At least the hunting of pigeons was for the purpose of eating. In killing the buffalo there was frequently no other purpose than a satisfaction in killing. In the twenty years after the Civil War, by appropriation of their habitat, we almost extinguished sixty to eighty million buffalo, the largest herds of hoofed animals ever known.

For transportation, we first dug an extensive network of canals, then drove our railways across the plains and through the mountains. All such feats were but predecessors of the time when we would pave over some of the most fruitful lands of the continent. This pavement was for millions of automobiles, racing to match the tempo at which life is lived, whether for trivial recreational pursuits or commercial exploitation.

Although severely damaged, the continent was able to survive our presence and our activities through most of the nineteenth century. But in the 1880s, a shift occurred that has determined the destiny of the American continent and planet Earth ever since. In that decade, we moved in a definitive way from an organic, renewable land-based economy to an extractive, nonrenewing industrial economy. This was a moment of immense consequence not only for ourselves but for the North American continent, for Western civilization, for the human community, and for the entire geobiological future of our world. Planet Earth was being turned from an ever-renewing to a nonrenewing way of life. We seem to have little understanding that a nonrenewing economy is a terminal economy, a one-way journey to extinction. Only an ever-renewing economy is a way of continuing life. Presently, as our national economy rises from millions to billions to trillions of dollars, as our oil tankers extend beyond a thousand feet in length, as we sink thousands of oil wells in every continent and every ocean, as we invade the rainforests of the planet and bring about the extinction of species at a rate unequalled since the last great geobiological extinction some sixty-five million years ago, as we melt the ice caps of the North and South Poles, as we upset the weather—as we observe all these events, we need to stop, think, and question.

How did we arrive at such a situation? This question lies hidden in the minds of those who ponder our future, although few people can articulate with any clarity just what has happened. When we came to this continent, we saw ourselves as a people with the most sublime spiritual insights in our biblical Christian heritage; as the most intellectual people of the world, with our great European universities of Oxford, Cambridge, Paris, Padua, Bologna, Louvain, and Berlin; as people with the most humane political traditions of the world, with our democratic political commitments; as people, through our technologies, most able to deal with the daily needs of the world for food, clothing, and shelter.

Now, after four centuries we find the North American continent toxic in its air, its water, and its land and gravely diminished in the variety and abundance of its living forms. We must ask ourselves: what happened? The answer is simply that we have lost our awareness that the human community exists only as a component of the larger Earth community. Instead of an intimate presence on an abundant continent that could inspire our minds and imaginations while providing for our physical needs, we became a predator people on an innocent continent. We withdrew from any intimate presence to this continent in order to give ourselves an identity, a dominance, and a destiny apart from the land. In reality, there is and can be no such thing as an enduring human community on this continent with a destiny apart from that of the continent itself. There is only one community on this continent, a single community that prospers or fails, that lives or dies together. The most foolish thing we can do is to think that we can enrich the human community by disrupting its integral functioning with the continental community. To damage this continent is to damage ourselves.

We should see our own countenance in the land we disfigure with our surface mining, with our dying forests, with our polluted rivers. We must wonder what gains justify the loss of such wonders as were freely given us by the natural context in which we found ourselves. We have become so conditioned to this contrived mode of living, this distorted use of natural resources, that even when we look for relief from the threatening impasse before us, we try to do so in a way that would not involve any real diminishment of the industrial lifestyle we have adopted. We continue to consume incessantly and blindly.

Apparently, during the four centuries since Descartes, we have lost our basic sensitivity for the ever-renewing natural world with its wonder, beauty, and intimacy, as well as its local and seasonal nourishment in response to our love and care of the land. We were willing to devastate all these for the illusory abundance offered by an industrial society.

Apparently, we needed no inspiration from the wonders of the natural world surrounding us. We received all that we needed from the Bible. Our religion fostered this alienation from the natural world by its neglect of the universe as our primary revelatory experience. In this perspective, the numinous dimension of the universe disappears completely. The majesty and mystery of the mountains, the rivers, the seas and sky, the forests with their trees and flowers, the meadows with their crickets and butterflies, the birds flying with the winds, the animals roaming the forests, and the oceans encompassing the continents have grown dim in human consciousness, their company no longer a personal or power presence.

What seems to be little understood is that our inner world of mind and imagination can only be activated by experience of the wonder and beauty of the outer world. If this outer world is damaged, there is progressive diminishment of our own personal fulfillment. We depend on the natural world in all its radiance to awaken in us our most precious intellectual, aesthetic, and emotional experiences. As humans, we could not have come into being until the natural world had achieved that brilliance of development characterizing the late Cenozoic period. We needed to experience a magnificent outer world to fulfill the needs of our inner world, our soul space.

The deep inner tendencies to dance and sing, the need to feel the wind in the summer evenings, to see the animals as they roam over the land: these awaken us to our personal identity and guide us in our fulfillment. Through what is seen in these surroundings we come to the knowledge of the unseen world of beauty beyond imagination, of intimacy with the numinous presence enfolding the entire universe. In the outer world of the universe we discover our complete self, our Great Self. This experience is what attracts a child running over the fields, touching everything, screaming with delight when first chasing a butterfly or playing with a young animal. In all these activities the child is discovering its own being.

We hardly know what we are doing when we adapt to our inventions. Even now we have barely a fragment of understanding of what we were doing when we began producing the automobile. We could not foresee and even now do not adequately understand the impasse we would experience in the lifestyle imposed by this vehicle. Nor could we imagine the effects on those basic processes of wind and weather, temperature and precipitation that would result. When we began the use of electric illumination in our cities, we did not understand or perhaps even care about the loss of our view of the stars at night. Yet this deprivation was the loss of an immense area in our psychic and emotional lives.

Further damage to ourselves as well as our surroundings began with the spread of artificial fertilizers, pesticides, and herbicides on the fields. We were naïve as to what we were doing to the entire biological community until Rachel Carson (1907–1964) wrote *Silent Spring* (1962), pointing out the devastation by pesticides of our entire biological surroundings. She pointed out that by poisoning the insects we were poisoning life forms that depended on the insects for food, especially birds. Her book was challenged, even condemned, even by scientists, as an unwarranted assault on the entire modern scientific world. Yet the consequences of DDT as she presented them would eventually be accepted as truth.

We could review specifics of the entire industrial devastation of the natural life systems, including the great dams on our western rivers: the Hoover Dam on the Colorado, sixty-four stories high; and the Glen Canyon Dam, 710 feet tall. The Grand Coulee Dam and the Bonneville on the Columbia River are colossal intrusions. In fact, we have built so many dams on the Columbia River that it has become something of a continuous reservoir for our electrical power, for our irrigation systems, and for the water supplying industrial and domestic demands. These dams, hundreds of feet tall, were built with utter disregard for the millions of salmon needing to return upstream to spawn. To accommodate the salmon at some of the dams, ladders were eventually built, some over sixty feet in height, but still, the salmon have to make an exhausting climb through falling water to reach their spawning areas. Young ones returning downstream must pass through hydroelectric generators that kill or seriously damage the vast majority on their first journey out to sea.

We were determined to reengineer the continent in all its basic functioning. Not satisfied with our physical, chemical, and electrical engineering, we even attempted to engineer humans, who needed to be restructured in their role of operating the all-encompassing system we had erected. One model of human engineering was promoted by Frederick Winslow Taylor (1856–1915). Through time studies of the best and most efficient manner in which humans could operate the machines now running the American economy, he devised mathematically efficient programs that integrated human functioning into machines that could not do their own thinking. Together the human operator and the machine made a single functioning unit. This adaptation of the human component to the machine was especially efficient in factory work. After he began his time studies in 1881, Taylor was hired by various corporations to assist in the effective engineering of humans to fit into various industrial processes. Henry Ford (1863–1947) applied these principles to create the assembly line, which he used to manufacture millions of automobiles.

Finally, in 1911, Taylor published *The Principles of Scientific Management*, which was enormously influential in the industrial and manufacturing establishments of the following decades. Throughout this new engineering development, our human role on Earth was altered from a meaningful presence on a living planet to that of a human component within a larger production process. While later management studies have adopted a more humanistic attitude, as exemplified in the work of Peter Drucker (1909–2005) and others like him, the earlier studies permanently and deleteriously marked the American industrial enterprise.

The frustrating element throughout this historical process is that the four establishments controlling our lives—government, corporations, universities, and religions—were all committed to or tolerant of unlimited exploitation of the entire range of planetary resources. Of the four, the economic-commercial establishment of corporations proved dominant. Government, for example, as we see in the present decisions of the U.S. Environmental Protection Agency, is extensively controlled by the economic powers. Universities serve the corporations as research centers for scientific studies, engineering training, and managerial skills. Likewise, many religious institutions have adopted corporate models of management and fundraising. Thus, their ambivalent position in the public life of the country is such that they can only feel awkward in critiquing the ills of the industrial way of life.

In some quarters, however, the deteriorating situation of the North American continent has at last evoked a rising tide of protest. Resentment at the actions of the industrial, commercial, and financial establishments has found expression in individual protests, organizational strategies, political activities, and educational reforms. Publications have been founded, studies have been done, and new ways of living in intimate association with the natural productivity of the land have been established.

Already in the nineteenth century, writers such as Henry David Thoreau (1817–1862) and John Muir (1838–1914) had begun to respond to the numinous dimension of the natural world. In the same period, Ralph Waldo Emerson (1803–1882) published his essay "Nature." These writers restored some of the ancient feeling of intimacy and personal fulfillment attained through presence to the forests and streams and meadows of Earth and to the stars and planets in the heavens.

We must remember, too, the inspiration provided by the European romantic movement as it emerged in the closing decades of the eighteenth century. Reacting against the Enlightenment initiated by Denis Diderot (1730–1784) and Jean le Rond D'Alembert (1717–1783), which emphasized intellectual developments and analytical thought, the romantic movement insisted on a more sensitive and imaginative view of human life and fulfillment. A major aspect of romanticism was its emphasis on the world of nature as found, for example, not only in the great English and European poets and novelists but in early nineteenth-century American writers such as William Cullen Bryant (1794–1878) and James Fenimore Cooper (1789–1851). In the same century, romantic devotion to nature led in America to the symbolic writings of Herman Melville (1819–1891), Nathaniel Hawthorne (1804–1864), and Edgar Allen Poe (1809–1849), and to the naturalism of Henry David Thoreau and Walt Whitman (1819–1892).

With its intensive focus on nature, this romantic age was also the period of the Hudson River painters, of Thomas Cole (1801–1848), Frederick Church (1826–1900), and Asher Brown Durand (1796–1886). Farther west, another remarkable series of painters emerged: George Catlin (1796–1892), Karl Bodner (1809–1893), and Albert Bierstadt (1830–1902), all recorders of early nineteenth-century Native American life.

These revelations in literature and in art of the imperatives of nature awakened others to the need for change. Individuals and some government agencies began to realize the extent of the environmental devastation. The conservation movement began to take shape. In 1872, Yellowstone National Park, an area of more than two million acres, was set aside by Congress to be preserved forever in its natural state. It was the first such national park to be established. In 1890, Yosemite National Park was endowed with similar provisions by Congress. In 1892, New York State set aside Adirondack State Park as a wilderness region in perpetuity. Although minimal in overall consequences for mitigating the effects of the industrial takeover of the continent, these enactments are examples of effective counteraction to the accelerating devastation of industrial society.

Private organizations, too, began to challenge predatory corporate activity. The Audubon Society, fostering the care of birds and other wildlife, was formed by George Grinnell (1849–1938) in 1886. John Muir founded the Sierra Club in 1894 as a means of protecting the natural world. In 1920, the Wilderness Society was established by Aldo Leopold (1887–1948). These initiatives were important beginnings of resistance to the continued attrition on natural life and habitat on this continent.

The main difficulty in restoring a viable mode of human presence on the land is that we have found ourselves so locked into and dependent on the existing industrial system for our food, shelter, clothing, transportation, and jobs that any thought of an alternative way of living has appeared more a dream than a possibility.

The paleontologist Fairfield Osborn (1857–1935) made it his mission to alert the people of this continent to the extent of the damage being done both here and throughout the world. His posthumous book *Our Plundered Planet* (1948) offered the first survey of what was happening to life systems throughout Earth. In 1954, Edward Solomon Hyams (1910–1975) wrote *Soil and Civilization*, a study of the causes of the decline of civilizations through the unlawful seizure of the commons and consequent impoverishment of the soil. The sharpest critique of the scientific and agricultural worlds was given in 1962 by Rachel Carson, in her study of DDT and its consequences for living creatures entitled *Silent Spring*. This work shocked America into awareness and even resentment never before experienced. It was a rude awakening.

A new environmental awareness began to appear in government conferences and institutions. In 1972, the first United Nations Conference on the Environment convened in Stockholm, Sweden. Some 90 percent of the countries represented subsequently established the first environmental protection agencies in their own countries. Also in 1972, Donella and Dennis Meadows's book *Limits to Growth* appeared, a study that has had immense influence. Millions of copies have been sold, and a thirty-year revised edition has been released. Then, in 1982, the United Nations produced a remarkable document entitled *A World Charter for Nature*, containing some of the finest statements yet made on environmental issues and the valuing of nature. Ironically, this document has been largely ignored by supposedly responsible authorities, news agencies, and even by environmentalists and ecologists.

Ten years later, in 1992, another United Nations Conference, this time in Rio de Janeiro, convened for the purpose of establishing global awareness of the need for sustainable development. The detailed outline of the program approved at the conference was contained in *Agenda 21*. This was the conference's main document, and it laid out a vision of sustainable development for the twenty-first century. Extended and detailed as it was, this item was so guarded against offending any of the political and economic powers that some feel it has become somewhat ineffective. One of the more encouraging developments coming from the Rio Earth Summit was the call for an Earth Charter, which was negotiated and drafted from 1997 through 2000. In three key sections—ecological integrity, social and economic justice, and democracy, nonviolence, and peace—the Earth Charter serves as an integrating ethical framework for a sustainable future for the planet. Moreover, in its preamble it contains an important reference to the large-scale evolutionary dimensions of our moment: "We are part of a vast, evolving universe. Earth, our home, is alive with a unique community of life."[4]

In addition to global governance, significant changes are being made in education. Among the most effective efforts are those of David Orr at Oberlin College in Ohio. Here, an ecology program has been established with a green building that exemplifies how humans can and should avail themselves of natural sources for heat, cooling, and electricity. There is also the work of Chet Bowers in Oregon, who has written extensively on education for a viable human presence on planet Earth. On a broader level, over 350 university presidents and chancellors representing more than forty countries have formally declared their commitment to environmental sustainability in higher education in a document known as the Talloires Declaration. This ongoing effort to "green" higher education has been led in this country by the association for University Leaders for a Sustainable Future (ULSF).

In economics, Herman Daly, Robert Costanza, and Richard Norgaard in 1990 established the International Society of Ecological Economics, which now has a membership from some forty different nations and publishes a journal distinguished for outlining the best strategies for shaping an economics appropriate for the century before us. In an effort to reduce the damage being done to Earth's biosystems by industrial processes, a program entitled "The Natural Step" was begun in Sweden in 1989 by Karl-Hendrik Robert. This movement proposed the diminishing of pressure against the natural world by using waste from workout processes for making new products. The proposal was taken up in America by Paul Hawken in *Natural Capitalism* (1997) and by Ray Anderson, who expands on the theory in his book *Midcourse Correction: Toward a Sustainable Enterprise* (1998). A book by Stephen Schmidheiny, *Changing Course: A Global Business Perspective on Development and the Environment* (1992), has also been helpful in this regard.

In agriculture, organic farms are being established, both by a large number of independent small farmers, such as Wendell Berry in Kentucky, and by farmers of larger enterprises, such as Fred Kirschenmann's two-thousand-acre farm in North Dakota. Over a thousand community-supported agricultural (CSA) projects are functioning throughout the country. Indeed, the list of ongoing agricultural initiatives in our national institutions is long. One of the most significant works has been that of Miriam Therese MacGillis, the founder of Genesis Farm in the Delaware Water Gap, an effort that is now thirty years old.

There are important efforts to engage the world's religions in rethinking human-Earth relations for a sustainable future. Foremost among these efforts is the religion and ecology project organized by Mary Evelyn Tucker and John Grim. This involved a series of ten conferences at Harvard's Center for the Study of World Religions from 1996 through 1998. This was followed by the publication of a series of ten books and the mounting of an international Web site for the Forum on Religion and Ecology. The forum is now located at Yale, where Tucker and Grim teach in a joint program between the School of Forestry and Environmental Studies and the Divinity School.[5]

Such listings as we have made of measures to alleviate abuses against the natural world and redirect human energies can serve as a background for ourselves as we venture into the twenty-first century. Yet none of these, or their aggregate, seems adequate for the revision we are suggesting. The continued devastation of the planet, the dysfunction of the life systems of Earth, the rate of extinction of living species involved in events such as the coming climate change: all these suggest the dimensions of the human-caused diminished health of Earth. The magnitude of the devastation demands more than the series of reforms we have cited. It calls for a change of mind, for universal acknowledgment that the human is a subdivision of Earth and as such bears a responsibility for Earth's health. It calls for global commitment to assist the planet in the recovery of its vigor. This view is promoted by some of the world's foremost biologists: E. O. Wilson of Harvard University, Peter Raven of the Missouri Botanical Garden, Paul Ehrlich of Stanford University, and Norman Myers of Cambridge University. The creation of that role, a global mission of regenerating healthy Earth-human relations, is our challenge. Our human presence on Earth must now make the transition from being a destructive force on the planet to being a pervasive life-giving presence.

First, we must identify in summary form the causes of our present situation. The first cause is our inability to understand that human beings find their fulfillment in the universe even as the universe finds its fulfillment in the human. The universe and planet Earth provide us with infinite wonder for our intelligence, entrancing beauty for our imagination, and profound presence for the healing of the human condition. The universe will forever remain the inspiration of our poetry, music, and arts. Without the outer world as it is, we would have no inner world of mind or imagination.

The second cause is our sense of Earth as primarily a natural resource for the unlimited use of humans. This sense of "use" of Earth as "natural resource" is an undermining aspect of our contemporary world and has pervaded Western civilization throughout its historical development. Its origin can be traced to the Jewish and Christian scriptures, where the presence of the divine in the cosmological order was diminished in favor of the divine as experienced in the historical order. This historical realism distinguishes Jewish and Christian scriptures from those of other civilizations. It is also that which has in certain ways defined Western religions, which tend to have a diminished sense of the divine in the cosmological order.

This spiritual overemphasis on the historical diminishes our appreciation of the natural world as the primary manifestation of an Originating Power enabling the universe to come into existence. To experience the universe and planet Earth as being primarily for use brings ruin to the natural world and destroys the poetry and the play that make life so fulfilling for humans.

The third cause of our difficulty here in North America is the U.S. Constitution, which establishes the doctrine of rights as an exclusive privilege of humans. If these rights for humans had been balanced with some assertion of the rights of the natural world or at least some protections, we might have kept ourselves from plundering the North American continent and much of the planet. This balancing, however, did not occur.

The fourth cause is the collaboration of the American legal profession and the judiciary with the commercial entrepreneurial enterprise in the economic development of the country. This development was noted by Morton J. Horwitz of Harvard University in his 1977 study *The Transformation of American Law: 1780–1860*:

> As political and economic power shifted to merchant and entrepreneurial groups in the post-Revolutionary period, they began to forge an alliance with the legal profession to advance their own interests through a transformation of the legal system. . . . By the middle of the nineteenth century the legal system had been reshaped to the advantage of men of commerce and industry at the expense of farmers, workers, consumers, and other less powerful groups within the society.[6]

This alliance was the beginning of the twentieth-century rise of the corporations as controlling forces in American society.

The commercial-industrial corporations have themselves grown profoundly aware of the damage resulting from their greed for monetary profit and controlling power. They have begun to recognize that they are bringing ruin on themselves as they ruin the planet on which they function. This awakening has led to the emphasis now given to "sustainable development" as a final norm of judgment in any economic situation.

Such is the beginning of an answer to the question of what has happened on the North American continent in the four centuries since the settlers from England landed along the southern seacoast of Virginia. In this survey of damage done, we are not proposing specific answers; we wish simply to clarify the conditions under which specific answers might be sought.

Even with projects such as the Swedish "Natural Step" or the Hawken and Lovins's book *Natural Capitalism*, the question we must ask is the extent to which we are willing to abandon our primary understanding of Earth as a natural resource for unlimited human use. We need instead to return to a primary understanding of Earth as that more profound dimension of our own existence, beyond pragmatic use, as the source whence we are born, the nourishment that sustains us while we are living, our healing in moments of distress, and the way to our final destiny.

Implementation of the proposals made in these and other movements is necessary if we are to establish a viable presence in the planetary community. We cannot do without the new technologies articulated in *Natural Capitalism* and shown to be coherent with the technologies of nature. Their adoption might be considered the first fundamental commitment needed by the contemporary industrial establishment. A further step in an acceptable program for the future is the resolution to shift from a view of the survival and benefit of the human as our primary concern to the survival and benefit of the comprehensive community of planet Earth as our primary concern.

Every individual self of Earth finds fulfillment in the Great Self of the Earth and beyond the Earth in the universe itself as the primary and ultimate expression of that mystery out of which all things come forth and then return in the full wonder of existence. In many of the proposals made so far, there is little more than increased concern for the details of the industrial-commercial order of our present lifestyle, more careful use of our resources, a diminishment of the pollution we are causing, and a reuse of waste. Beyond these details lies the larger issue: recovery of the distinctive economy of the organic world, which offers an ever-renewing process leading through the rise and dissolution of life to give new life to the succeeding generation. Combined with this principle is a need for greater dependence on energy resources such as sunlight and wind and water, which are not exhausted by use.

Pragmatic efforts at establishing a viable way into the future are urgently needed and invaluable. They are indispensable in any effort to deal with that future. Even with the change in attitude that I am proposing, the details of implementation will be an essential aspect of any future program. I do not wish to diminish what is being done. I wish only to indicate that the basic difficulty lies deeper in the human mind and emotions than is generally recognized. If the reorientation of mind is not effected, then whatever remedy is proposed will not succeed in the purposes it intends.

So far, we have not been able to effect a major change of inner attitude that would enable us to return from our extractive, nonrenewing, industrial way of life to an organic, ever-renewing, land-based way of life. We seem not to recognize the psychic addictions that we have acquired through these last centuries, when we thought that we were moving from a primitive, less human status to a truly human way of life based on our scientific-technological dominion over Earth and its resources. Earth henceforth would serve us. We thought we had established ourselves beyond the controls and limitations of Earth's natural systems. But that wonderful interplay between ourselves and those natural forces of the Earth experienced as ever-renewing presences is what needs to be fostered more than ever before.

CHAPTER 13

The World of Wonder

WHAT DO you see? What do you see when you look up at the sky at night at the blazing stars against the midnight heavens? What do you see when the dawn breaks over the eastern horizon? What are your thoughts in the fading days of summer as the birds depart on their southward journey, or in the autumn when the leaves turn brown and are blown away? What are your thoughts when you look out over the ocean in the evening? What do you see?

Many earlier peoples saw in these natural phenomena a world beyond ephemeral appearance, an abiding world, a world imaged forth in the wonders of the sun and clouds by day and the stars and planets by night, a world that enfolded the human in some profound manner. This other world was guardian, teacher, healer—the source from which humans were born, nourished, protected, guided, and the destiny to which we returned.

Above all, this world provided the psychic power we humans needed in our moments of crisis. Together with the visible world and the cosmic world, the human world formed a meaningful threefold community of existence. This was most clearly expressed in Confucian thought, where the human was seen as part of a triad with Heaven and Earth. This cosmic world consisted of powers that were dealt with as persons in relationship with the human world. Rituals were established whereby humans could communicate with one another and with the earthly and cosmological powers. Together these formed a single integral community—a universe.

Humans positioned themselves at the center of this universe. Because humans have understood that the universe is centered everywhere, this personal centering could occur anywhere. For example, the native peoples of North America offered the sacred pipe to the powers of the four directions to establish themselves in a sacred space where they entered into a conscious presence with these powers. They would consult the powers for guidance in the hunt, strength in wartime, healing in time of illness, support in decision making. We see this awareness of a relationship between the human and the powers of the universe expressed in other cultures as well. In India, China, Greece, Egypt, and Rome, pillars were established to delineate a sacred center, which provided a point of reference for human affairs and bound Heaven and Earth together.

There were other rituals whereby human communities validated themselves by seasonal acknowledgement of the various powers of the universe. This is still evident with the Iroquois autumn thanksgiving ceremony, where the sun, the Earth, the winds, the waters, the trees, and the animals each in turn received expressions of personal gratitude for those gifts that made life possible. Clearly, these peoples see something different from what we see.

We have lost our connection to this other deeper reality of things. Consequently, we now find ourselves on a devastated continent where nothing is holy, nothing is sacred. We no longer have a world of inherent value, no world of wonder, no untouched, unspoiled, unused world. We think we have understood everything. But we have not. We have *used* everything. By "developing" the planet, we have been reducing Earth to a new type of barrenness. Scientists are telling us that we are in the midst of the sixth extinction period in Earth's history. No such extinction of living forms has occurred since the extinction of the dinosaurs some sixty-five million years ago.

There is now a single issue before us: survival. Not merely physical survival, but survival in a world of fulfillment, survival in a living world, where the violets bloom in the springtime, where the stars shine down in all their mystery, survival in a world of meaning. All other issues dwindle in significance—whether in law, governance, religion, education, economics, medicine, science, or the arts. These are all in disarray because we told ourselves: We know! We understand! We see! In reality what we see, as did our ancestors on this land, is a continent available for exploitation.

When we first arrived on this continent some four centuries ago, we also saw a land where we could escape the monarchical governments of Europe and their world of royalty and subservience. Here before us was a land of abundance, a land where we could own property to use as we wished. As we became free from being ruled over, we became rulers over everything else. We saw the white pine forests of New England, trees six feet in diameter, as forests ready to be transformed into lumber. We saw meadowland for cultivation and rivers full of countless fish. We saw a continent awaiting exploitation by the chosen people of the world.

When we first arrived as settlers, we saw ourselves as the most religious of peoples, as the most free in our political traditions, the most learned in our universities, the most competent in our technologies, and most prepared to exploit every economic advantage. We saw ourselves as a divine blessing for this continent. In reality, we were a predator people on an innocent continent.

When we think of America's sense of "manifest destiny," we might wish that some sage advice regarding our true role had been given to those Europeans who first arrived on these shores. We might wish that some guidance in becoming a life-enhancing species had been offered during these past four centuries. When we first arrived on the shores of this continent, we had a unique opportunity to adjust ourselves and the entire course of Western civilization to a more integral presence to this continent.

Instead, we followed the advice of the Enlightenment philosophers, who urged the control of nature: Francis Bacon (1561–1626), who saw human labor as the only way to give value to the land; Rene Descartes (1596–1650); and John Locke (1632–1704), who promoted the separation of the conscious self from the world of matter. In 1776, when we proclaimed our Declaration of Independence, we took the advice of Adam Smith's (1723–1790) *Inquiry Into the Nature and Causes of the Wealth of Nations*, a book of enormous influence in the world of economics from then until now. Our political independence provided an ideal context for economic dominance over the natural world.

As heirs to the biblical tradition, we believed that the planet belonged to us. We never understood that this continent had its own laws that needed to be obeyed and its own revelatory experience that needed to be understood. We have only recently considered the great community of life here. We still do not feel that we should obey the primordial laws governing this continent, that we should revere every living creature—from the lowliest insect to the great eagle in the sky. We fail to recognize our obligation to bow before the majesty of the mountains and rivers, the forests, the grasslands, the deserts, the coastlands.

The indigenous peoples of this continent tried to teach us the value of the land, but unfortunately we could not understand them, blinded as we were by our dream of manifest destiny. Instead we were scandalized, because they insisted on living simply rather than working industriously. We desired to teach them our ways, never thinking that they could teach us theirs. Although we constantly depended on the peoples living here to guide us in establishing our settlements, we never saw ourselves as entering into a sacred land, a sacred space. We never experienced this land as they did—as a living presence not primarily to be used but to be revered and communed with.

René Descartes taught us that there was no living principle in the singing of the wood thrush or the loping gait of the wolf or the mother bear cuddling her young. There was no living principle in the peregrine falcon as it soared through the vast spaces of the heavens. There was nothing to be communed with, nothing to be revered. The honeybee was only a mechanism that gathered nectar in the flower and transformed it into honey for the sustenance of the hive and the maple tree only a means of delivering sap. In the words of a renowned scientist: "For all our imagination, fecundity, and power, we are no more than communities of bacteria, modular manifestations of the nucleated cell."[1]

In order to counter reductionistic and mechanistic views of the universe such as this, we need to recover our vision, our ability to see. In the opening paragraph of *The Human Phenomenon*, Pierre Teilhard de Chardin (1881–1955) tells us: "One could say that the whole of life lies in seeing. That is probably why the history of the living world can be reduced to the elaboration of ever more perfect eyes. . . . See or perish. This is the situation imposed on every element of the universe by the mysterious gift of existence."[2] We need to begin to see the whole of this land. To see this continent, we might imagine ourselves in the great central valley that lies between the Appalachian Mountains to the east and the Rocky Mountains to the west. Here we would be amazed at the vast Mississippi River, which flows down through this valley and then on into the immense gulf that borders the southern shores of this continent. This massive flow of water, including its tributary the Missouri, flowing in from the northwest, constitutes one of the greatest river systems on the planet, draining almost the entire continent, from New York and the Appalachian Mountains in the east to Montana and the Rocky Mountains in the west.

This region includes the Great Plains, the tall grasslands that extend from Indiana to the Mississippi River, to the short grasslands that begin across the river and extend to the mountains. This is a territory to be honored in some special manner. The region to the west of the river has what are among the deepest and most fertile soils on the planet. Soils that elsewhere are only inches in depth here are several feet deep, soils formed of the debris washed down from the mountains over the long centuries. A large human population depends on this region. Such precious soil is a gift to be carefully tended. This center of commercial wheat, and later corn production, began in New York in the early nineteenth century and extended westward until now it can be located in those Kansas fields of grain that extend beyond the horizon.

When we stand in the Mississippi basin, we can turn westward and experience the mystery, adventure, and promise of this continent; we can turn eastward and feel its history, political dominance, and commercial concerns. Westward are the soaring redwoods, the sequoia, the Douglas fir, the lodgepole pine; eastward are the oaks, the beech, the sycamore, the maple, the spruce, the tulip poplar, the hemlock. Together, these bear witness to the wonder of the continent and the all-encompassing sea.

We might also go to the desert, or high in the mountains, or to the seashores, where we might really see, perhaps for the first time, the dawn appear in the eastern sky—its first faint purple glow spreading over the horizon, then the slow emergence of the great golden sphere. In the evening, we might see the flaming sunset in the west. We might see the stars come down from the distant heavens and present themselves almost within reach of our arms if we stood on tiptoe.

So too, we might begin to view the change of the seasons: the springtime awakening of the land as the daisies bloom in the meadows and the dogwood tree puts forth its frail white blossoms. We might experience the terrifying moments when summer storms break over the horizon and lightning streaks across the sky, the moments when darkness envelops us in the deep woodlands, or when we experience the world about us as a vast array of powers asserting themselves. When we view all this, we might begin to imagine our way into the future.

Concerning this future we might make two observations. First, the planet Earth is a onetime project. There is no real second chance. Much can be healed because the planet has extensive, albeit limited, powers of recovery. The North American continent will never again be what it once was. The manner in which we have devastated the continent has never before occurred. In prior extinctions, the land itself remained capable of transformations, but these are now much more difficult to effect. Second, we have so intruded ourselves and debilitated the continent in its primordial powers that it can no longer proceed simply on its own. We must be involved in the future of the continent in some comprehensive manner.

It is clear that there will be little development of life here in the future if we do not protect and foster the living forms of this continent. To do this, a change must occur deep in our souls. We need our technologies, but this is beyond technology. Our technologies have betrayed us. This is a numinous venture, a work of the wilderness. We need a transformation such as the conservationist Aldo Leopold (1887–1948) experienced when he saw the dying fire in the eyes of a wolf he had shot. From that time on, he began to see the devastation that we were bringing upon this continent. We need to awaken, as did Leopold, to the wilderness itself as a source of a new vitality for its own existence. For it is the wild that is creative. As we are told by Henry David Thoreau (1817–1862), "In Wildness is the preservation of the world."[3] The communion that comes through these experiences of the wild, where we sense something present and daunting, stunning in its beauty, is beyond comprehension in its reality, but it points to the holy, the sacred.

The universe is the supreme manifestation of the sacred. This notion is fundamental to establishing a cosmos, an intelligible manner of understanding the universe or even any part of the universe. That is why the story of the origin of things was experienced as a supremely nourishing principle, as a primordial maternal principle, or as the Great Mother, in the earliest phases of human consciousness. Some of the indigenous peoples of this country experience it as the Corn Mother or as Spider Woman. Those who revere the Corn Mother place an ear of corn with the infant in the cradle to provide for the soothing and security the infant needs to feel deep in its being. From the moment the infant emerges from the warmth and security of the womb into the chill and changing world of life, the ear of corn is a sacred presence, a blessing.

We must remember that it is not only the human world that is held securely in this sacred enfoldment but the entire planet. We need this security, this presence throughout our lives. The sacred is that which evokes the depths of wonder. We may know some things, but really we know only the shadow of things. We go to the sea at night and stand along the shore. We listen to the urgent roll of the waves reaching ever higher until they reach their limits and can go no farther, then return to an inward peace until the moon calls again for their presence on these shores.

So it is with a fulfilling vision that we may attain—for a brief moment. Then it is gone, only to return again in the deepening awareness of a presence that holds all things together.

Notes

2 Religion in the Global Human Community

1 Mircea Eliade, *The Quest: History and Meaning in Religion* (Chicago: University Of Chicago Press, 1984), 62.

2 Ibid., 59.

3 René Dubos, *The Dreams of Reason: Science and Utopias* (New York: Columbia University Press, 1961), 123.

3 Alienation

1 *Yoga Sutras of Patanjali* 1.3.

2 Mencius 4.2.19.

3 Mencius 6.1.11.

4 Historical and Contemporary Spirituality

1 John Neihardt, *Black Elk Speaks: Being the Life Story of a Holy Man of the Oglala Sioux* (New York: Pocket Books, 1972), 35.

6 Religion in the Twenty-first Century

1 Peter Raven, "We're Killing Our World: The Global Ecosystem in Crisis," Keynote speech, American Association of the Advancement of Science, Chicago, 1987.

7 Religion in the Ecozoic Era

1 Charles Krauthammer, "Essay," *Time* (June 17, 1991).

2 J. E. Strickland, ed., *William Strickland's Journal of a Tour in the United States* (New York: New-York Historical Society, 1971).

3 Col. 1:17.

4 Wang Yang-ming, "Inquiry on the Great Learning," in *Instructions for Practical Living and Other Neo-Confucian Writings by Wang Yang-ming*, trans. Wing-tsit Chan (New York: Columbia University Press, 1963), 273.

8 The Gaia Hypothesis: Its Religious Implications

1 H. Frankfort, H. A. Frankfort, John A. Wilson, Thorkild Jacobsen, and William A. Irwin, *The Intellectual Adventure of Ancient Man: An Essay on Speculative Thought in the Ancient Near East* (Chicago: The University of Chicago Press, 1946), 4.

2 Interview with Edgar Mitchell by Stanley Rosen, Palo Alto, Calif., July 1974. Quoted in Kevin W. Kelley, *The Home Planet* (Reading, Mass.: Addison-Wesley, 1988), 138.

3 Thomas Aquinas *Summa Theologica* 1.47.l.

9 The Cosmology of Religions

1 Thomas Aquinas *Summa Contra Gentiles* 2.45.10.

10 An Ecologically Sensitive Spirituality

1 The Forum on Religion and Ecology at Yale's Web site is located at http://fore.research.yale.edu

11 The Universe as Divine Manifestation

1 David Abram, *The Spell of the Sensuous: Perception and Language in a More-Than-Human World* (New York: Vintage Books, 1997), 264.

12 The Sacred Universe

1 The statement by Thomas Paine is found in the appendix to the third edition of *Common Sense*: "A situation, similar to the present, hath not happened since the days of Noah until now."

2 Annette Kolodny, *Lay of the Land: Metaphor as Experience in American Life and Letters* (Chapel Hill: University of North Carolina Press, 1975), 10.
3 Speech by Powhatan, as recorded by John Smith, 1609, in P. L. Barbour, ed., *The Jamestown Voyages*, 375, and quoted in T. C. McLuhan, *Touch the Earth: A Self-Portrait of Indian Existence* (Edison, N.J.: BBS Publishing Corp., 1992), 66.
4 http://www.earthcharter.org.
5 http://fore.research.yale.edu.
6 Morton J. Horwitz, *The Transformation of American Law: 1780–1860* (Cambridge, Mass.: Harvard University Press, 1979), 253–254.

13 The World of Wonder

1 Lynn Margulis and Dorion Sagan, *Microcosmos: Four Billion Years of Microbial Evolution* (Berkeley: University of California Press, 1997), 191.
2 Pierre Teilhard de Chardin, *The Human Phenomenon*, trans. Sarah Appleton-Weber (Eastbourne, East Sussex: Sussex Academic Press, 1999), 3.
3 Henry David Thoreau, *Walking: A Little Book of Wisdom* (New York: Harper Collins, 1994), 19. This was originally published as the essay "Walking" in *Atlantic Monthly*, after Thoreau's death in 1862.